Essential Principles

OF

Image Sensors

Essential Principles

OF

Image Sensors

TAKAO KURODA

CRC Press
Taylor & Francis Group
Boca Raton London New York

CRC Press is an imprint of the
Taylor & Francis Group, an **informa** business

CRC Press
Taylor & Francis Group
6000 Broken Sound Parkway NW, Suite 300
Boca Raton, FL 33487-2742

First issued in paperback 2017

Version Date: 20140319

ISBN 13: 978-1-4822-2005-6 (hbk)
ISBN 13: 978-1-138-07417-0 (pbk)

Library of Congress Cataloging-in-Publication Data

Kuroda, Takao.
  Essential principles of image sensors / Takao Kuroda.
    pages cm
  Summary: "Providing an introduction to the systemization of image sensor technology and perspective on its various noise sources and signal processes, this book delivers a detailed description of image information and its four factors: light intensity, space, wavelength, and time. It discusses how image sensors convert optical image information into image signals. It tackles CCD, MOS, and CMOS, as well as key techniques such as BSI. It also considers the influences of in-system digitized coordinate points and explains sampling theorem, presenting unique figures demonstrating the importance of phase"-- Provided by publisher.
    Includes bibliographical references and index.
    ISBN 978-1-4822-2005-6 (hardback)
    1. Image converters. I. Title.

TK8316.K87 2014
681'.25--dc23   2014009743

Visit the Taylor & Francis Web site at
http://www.taylorandfrancis.com

and the CRC Press Web site at
http://www.crcpress.com

Contents

Preface to English Edition

It is my greatest pleasure that this book concerning image sensors will be available for people to read all over the world.

The Japanese edition of this book was published in December 2012. Its aim was to explain the "indispensable functions and ideal situation of image sensors" by marshaling the "factors defining image information quality" and understanding the "structure of image information." It also discussed the history of image sensors and the way they tried to respond to market demands by using the best combinations of the available technology in each era. The book also attempted to investigate future directions of this field.

Although this book was started only 5 months after the publication of the Japanese edition, there was significant progress in this short time, and these advances are described in this edition. While the content of this book is "essential principles," as the title shows, it also deals with the "state of the art" regarding the surrounding issues.

This edition has come to be published thanks to Professor Dr. Jun Ohta of the Nara Institute of Science and Technology, who kindly introduced me to CRC Press/Taylor & Francis. The author genuinely expresses his gratitude. Thanks are extended to the relevant staff of CRC Press/Taylor & Francis for their hearty support.

I hope that this book enables readers to survey the whole world of image sensor technology through an understanding of the essential principles.

Finally, I gratefully acknowledge the support of my family. My wife, Toshiko, established and has been maintaining the circumstances to enable me to devote myself wholeheartedly to my job for 35 years and checked the English throughout the book. There is little doubt that this book, including the Japanese edition, would not have been born without her support. I thank my two daughters, Chiaki and Chihiro, who encouraged me and helped with the English and some of the artwork from overseas. I also thank my dog, Simba, who was a treasured member of our family and appears in many of the sample images throughout this book. She healed and heartened me with her gentle and friendly nature, charming and cute smile, elegant appearance, and amazing intelligence.

Takao Kuroda
Ibaraki, Osaka, Japan

Preface to Japanese Edition

More than thirty years have passed since I entered the field of image sensor technology. During this time, my main focus has been how image sensors should be understood. During these years, I have finally achieved a solution. In a sense, this book is also a reminder to sort out my remaining tasks. Thus, I ask readers for permission to present many of my own unique ways of understanding, explanations, structures of logics, and sometimes biases.

At university, I studied physics, which favors a unified perspective, and I believe that to understand is to be able to call up an image of the phenomenon in the mind. To realize my belief, I have tried to visualize physical phenomena by using my original figures to help readers' intuitive understanding.

Chapter 1 presents the structure of image information and the role of sensors to make it clear what "imaging" is, as the basis of this book. In Chapter 2, semiconductor components and circuit components, necessary for sensors, will be dealt with. In Chapter 3, the noise in sensors will be described. In Chapter 4, the scanning mode will be dealt with to conclude the preparatory stages.

Chapter 5 contains a detailed explanation of the principles, pixel technologies, and progress of CCD, MOS, and CMOS sensors. Their current situations will be compared at the end of the chapter. Chapter 6 contains an explanation of how information other than light intensity is obtained. In Chapter 7, the techniques that have improved information quality of each factor constructing the image will be described. In Chapter 8, the elements concerning the information quality of images, not of single sensors but of whole imaging systems, will be described.

In publishing this book, I am indebted to many people in this field, including my elders and seniors, some of whom have already passed away. And in my office, I was so happy to have respectable bosses, nice seniors, colleagues, and juniors, including a few people who showed examples of how not to behave. I would especially like to sincerely thank Dr. Kenju Horii, who was my first direct boss in the office and guided me from the very beginning. My work would not have continued to today without meeting Dr. Horii. I found my true vocation by providence. I also would like to express special thanks to Mr. Yoshiyuki Matsunaga, who was seated beside me during my last years in the office. We had many discussions as engineers of image sensors and also as humans. I learned much from his words and behavior.

As I wrote this book at home, I am very grateful for the constant support of my family, wife, two daughters, and dog, Simba, who often appears in this book in the sample images and was so gentle, generous, and smart. She always cheered me up, but sadly she passed away while I was writing the final chapter of the book.

In publishing this book, I specially acknowledge Professor Dr. Takayuki Hamamoto of Tokyo University of Science for his kind efforts and the relevant members of the Corona Publishing Company. I would also like to thank the authors whose figures from papers and websites I have referred to. As for the relevant literature on the fundamental techniques, I have tried to refer to the most original works or those published earliest.

If this book is of help in guiding readers who want to understand the world of image sensors but are at a loss, as I was in younger days, I will be most grateful.

Takao Kuroda
Ibaraki, Osaka, Japan

Author

Takao Kuroda received his bachelor and master's degrees in material physics and PhD from Osaka University, Osaka, Japan, in 1972, 1974, and 1978, respectively.

He joined Panasonic Corporation in Osaka in 1978 and started research and development of image sensors, beginning with charge-coupled devices, at the Panasonic Electronic Laboratory. He extended the field of development to shift-register addressing-type sensors in addition to charge-coupled devices. In 1981, a sensor model he designed, a solid-state single-chip color camera system, was employed for the cockpit cameras of Japan Airlines to show views at takeoff and landing to passengers for the first time. In 1985 he was in charge of evaluation in the development of the first mass-produced charged-couple device model for Panasonic's single-chip consumer color camcorder. In 1986 he developed the first charged-couple device model for consumer camcorders that featured an electronic shutter function. He also developed a very-low-smear-level charged-couple device with a new photodiode structure, which was reported at the International Solid-State Circuits Conference in 1986.

In 1987, he moved to Panasonic's Kyoto Research Laboratory to establish a charged-couple device developmental regime and processing line. In 1989, he happened to encounter high-energy ion implantation technology, which he instantly recognized as having remarkable potential for performance improvement of image sensors, especially for charged-couple devices. He had a clear view of their absolute necessity, as present conditions prove. After confirmation of the performance improvement of a prototype charged-couple device, he installed the equipment in the laboratory. He was the manager of the elemental technology development group.

He was developing image sensors considering the essentiality of imaging and future integration with computers and communication. As a result, he and his group developed a technique that could remarkably increase the maximum charge quantity transferred in charged-couple devices, as reported at the International Solid-State Circuits Conference in 1996. This brought about significant performance advances in all Panasonic charged-couple devices, some of which were used in camcorders by a competitor charged-couple device company.

As he had recognized the potentiality of combining complementary metal-oxide semiconductor sensors with the pixel technology of charged-couple devices, he embarked upon this project for the first time at Panasonic in 1996, despite much dissent.

In 1998, he moved to the image sensor business unit to establish a business strategy, while addressing the issues of the charged-couple device mass-production line, which had shifted to the Tonami factory, and developing a complementary metal-oxide semiconductor sensor. He was in charge of reinforcement of the development system and intellectual property from 2001 to 2005.

Initially he retired from Panasonic in December 2005, but at the request of the company in January 2006, he took up a post as advisor to the image sensor business unit to establish a technology strategy based on his views on technology and business.

He retired completely from Panasonic in 2011 to write a book on image sensor technology. The Japanese edition was published in 2012 by Corona Publishing Co., Ltd.

Dr. Kuroda was a member of the subcommittee on imagers, microelectromechanical systems, and medical and display devices for the International Solid-State Circuits Conference from 2006 to 2008. He is a fellow of the Institute of Image Information and Television Engineers of Japan.

He holds 70 Japan patents and 15 U.S. patents.

1

Task of Imaging and Role of Image Sensors

Various kinds of instruments such as digital still cameras (DSCs), camera phones, and camcorders allow us to enjoy personal images. Moreover, broadcasting cameras, which provide high-definition images, are indispensable to the television industry. Not only are there cameras for personal enjoyment, but there are also cameras for other applications such as for automobiles and security systems, and endoscopes for medical use. Cameras are not only used for visible light, but they are also used for thermography, which visualizes thermal distribution by infrared imaging; there are also cameras for ultraviolet and x-ray imaging. In addition, there are cameras that capture very-high-speed phenomena and cameras that obtain highly accurate color information. Furthermore, there are cameras whose images are used not by the human eye but by machines, such as cameras for automated driving and machine vision, which judge information obtained from images. As just described, various kinds of cameras are utilized in a very wide range of fields.

Why are there so many kinds of imaging systems as typified by cameras? The reason is to obtain images with adequate image quality for the purpose of each imaging system. In each imaging system, the role of each image sensor is to pick up image information of high enough quality for that system.

In this chapter, the factors that determine image information are confirmed. Then, the structure of image sensor output and image information are set out. Unless explicitly stated otherwise, the explanations are based on "almost all image sensors" (see Section 1.2.3). Concerning the terms used in this book, *image information* is used for image information in a broader sense, *optical image information* is used for information contained in optical images, and *image signal* is used for image sensor output obtained from optical images.

1.1 Factors Constructing Image Information

What is image information made up of? For the sake of simplicity, let us initially focus on a monochrome still picture. There is a concentration distribution of black and white in two-dimensional space in a monochrome still picture. The concentration indicates the light intensity, which is brighter at lower concentrations. That is, the concentration is the light intensity distribution at each position in two-dimensional space. Therefore, an image is constructed using the information on space (position) and the intensity of the light at that position.

Let us now consider color still images. Since information on the wavelength of light must be added, color image information is formed by light intensity, space, and wavelength. Moreover, in the case of color moving images, time information when light reaches the image should be added. Thus, the image information is composed of four factors: light intensity, space (position), wavelength, and time.[1]

Among these factors, space has two dimensions, however wavelength information is often replaced and approximated by the primary colors red, green, and blue, as will be shown later; therefore, color can be considered three dimensional and time has one dimension. Thus, as shown in Figure 1.1, a distribution of the four factors in seven dimensions, that is, a set of each of the coordinate points of light intensity, space, wavelength, and time constructs the image information.

The indexes that indicate the level of information quality are accuracy and range, as shown in Figure 1.2. While the accuracy of value is the resolution capability, which means the signal-to-noise ratio (SNR), the range is the extent of the signal information that the imaging system can pick up. Information captured with high accuracy and over a wide range is high-quality image information. In the case of light intensity, for example, the information quality is decided by the level of SNR and the dynamic range, which decribes the maximum and minimum measurable light intensities that the imaging system can capture, as shown in Figure 1.3.

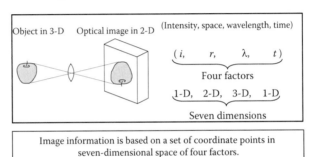

FIGURE 1.1
Structure of optical image information.

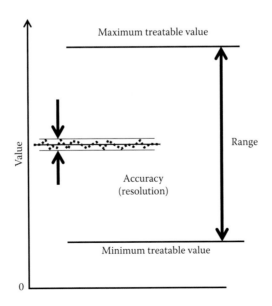

FIGURE 1.2
Quality of image information: Accuracy and range.

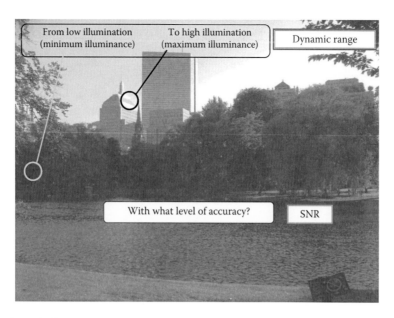

FIGURE 1.3 **(See color insert)**
Example of the quality of light intensity information.

The accuracy and range of the four factors are shown in Table 1.1. The accuracy of space information is space resolution, that of wavelength is color reproducibility, and that of time is time resolution. The ranges of space,* wavelength, and time are the capturing space range, the color gamut or wavelength range, and the storage time range,† respectively. Apart from special applications and the dynamic range of intensity, these ranges are rarely a problem.

In addition to intensity and wavelength, polarization and phase are among the light conditions. Although these are not discussed in this book, some examples deal with this information, such as sensors to obtain the polarization distribution,[2] which concerns the surface conditions of materials and textured surfaces, and sensors[3] and systems[4] to obtain depth and range information using phase difference.[5] There is a sensor[6] that gets range information from the angle information of incident light. Since there is a strong need for information on range and depth, progress in this field is expected.

TABLE 1.1

Accuracy and Range of Four Factors

Factor	Accuracy (Resolution)	Range
Intensity	SNR (sensitivity)	Dynamic range
Space	Space resolution	Space range
Wavelength	Color reproducibility	Color gamut, wavelength range
Time	Time resolution	Storage time range

---

* Although space is directly related to the sensor size, it can be extended by using a wide lens such as a fish-eye lens.
† Storage memory size.

1.2 Image Sensor Output and Structure of Image Signal

In this section, the structure of an image signal captured by image sensors is discussed.

1.2.1 Monochrome Still Images

As discussed in Section 1.1, the only image information required for monochrome still images is light intensity and space (position). The basic device configuration of image sensors is shown in Figure 1.4. Image sensors have an *image area* on which optical images are focused and are converted to image signals for output. In the image area, unit cells called pixels are arranged in a matrix in a plane. Each pixel has a sensor part typified by a photodiode, which absorbs incident light to generate a certain quantity of signal charges according to the light intensity. Thus, the light intensity information for a pixel is obtained at each sensor part. Figure 1.5b shows a partially enlarged image of Figure 1.5a, and the image signal of the same area captured by image sensors is shown in Figure 1.5c. The density of each rectangular block is the output of each pixel and the light intensity information $S(x_i, y_j)$ at the coordinate point (x_i, y_j) of each pixel. Expressing a two-dimensional coordinate point (x_i, y_j) using r_k, the signal can be written as $S(r_k)$. A set of $S(r_k)$ at all r_k in the image area makes up the information for one monochrome still image.

Dividing the image area into pixels that have a finite area, as shown in Figure 1.6, means fixing the area size and coordinate point at which light intensity information is picked up. The figure also shows that the continuous analog quantity, representing the position information, is replaced by discrete coordinate points. That is, x_i and y_j cannot take arbitrary values, but are built-in coordinate points that are fixed in imaging systems, which are the digitization of space coordinates. Using digitization in this system, since two-dimensional space information can be treated as determinate coordinate points, only light intensity information should be obtained. Therefore, the three-dimensional information of a monochrome still image is compressed into one dimension. Since the overall total

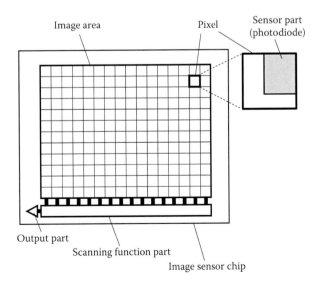

FIGURE 1.4
Basic device configuration of image sensors.

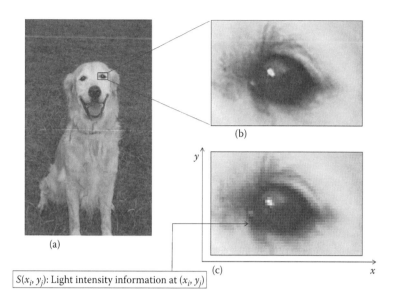

$S(x_i, y_j)$: Light intensity information at (x_i, y_j)

FIGURE 1.5 (See color insert)
Optical image and the image signal obtained by image sensors: (a) optical image; (b) enlarged optical image; (c) image signal obtained by sensors.

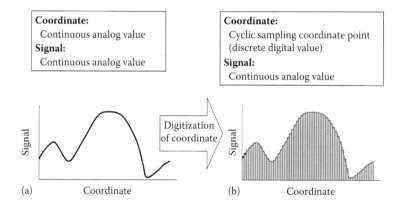

FIGURE 1.6
Digitization of coordinate: (a) space coordinate of an optical image; (b) space coordinate in an image sensor.

number of coordinate points is a pixel count of the image sensors, a larger number of pixels can achieve a higher space frequency, that is, a higher space resolution.

Image sensors also have a scanning function, transmitting the light intensity information of each pixel to the output part, which reads out the intensity information as an electric signal, as shown in Figure 1.4.

1.2.2 Color Still Images

The image information of color still images is formed by light intensity, space, and wavelength, as discussed in Section 1.1. Therefore, it is necessary to take the wavelength information in addition to the monochrome still image information. Although there are several ways to gather wavelength information, the most common method, the single-sensor color system in which only one image sensor is used, is described. In this system, a color filter array (CFA) is arranged on a sensor. Figure 1.7 shows the Bayer CFA arrangement,[7] which is by far the most commonly used CFA, with a pitch formed by 2 × 2 pixels. The pitch is composed of the three primary colors: red (R), green (G), and blue (B). On each pixel, only one color part of the filter is arranged. That is, one pixel for color imaging is formed from each one-color filter and the sensing part is formed under it. Since the human eye has the highest sensitivity to green, the G filter is arrayed in a checkerboard pattern* to contribute most to the space resolution. Red and blue are arranged between green on every two lines.

Using this configuration, each pixel contains information from only one color among R, G, and B. Although the wavelength distribution is continuous and analog, color information is restricted to only three types. Since the identification of a pixel also means the identification of its color, wavelength information also contains built-in coordinate points of color in the system as well as space information. By showing a coordinate point (R, G, or B) in three-dimensional space as c, the output from pixel k and color l can be expressed as $S(r_k, c_l)$. A set of signals from each pixel and color coordinate constructs the color still image information.

As a pixel and its color are specified at the same time in a single-sensor color system, k and l are immediately identified. The information on the physical quantity "wavelength" is replaced by perception "color." In this system, the perceptual mechanism of color by the human eye and brain is availed of, and will be discussed in detail in Section 6.4.

1.2.3 Color Moving Images

The image information of color moving images requires all four factors (light intensity, space, wavelength, and time), as previously discussed. Almost all moving images are achieved by the continuous repetitive capturing and reproducing of still pictures, as shown in Figure 1.8. Thus, the basis of moving images is still images. Despite the name "still" images, their light intensity information is not that of an infinitesimal time length, but integrated information gathered during an exposure period of a certain length. This operation mode is called the integration mode. In this way, the signal amount can be increased by integrating the signal charges being generated throughout the exposure period. Since sensitivity can be drastically improved, almost all image sensors employ the integration mode.[†]

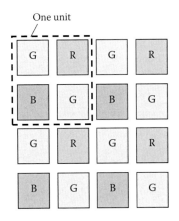

FIGURE 1.7
Example of the digitization of wavelength information: Bayer color filter array arrangement.

---

* The relationship between the arrays will be discussed in Section 7.2.
† "Almost all image sensors" means the sensors with integration mode.

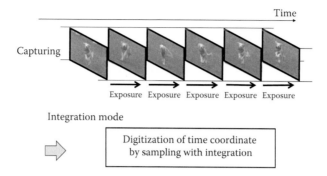

FIGURE 1.8
Capturing the sequence of moving pictures.

In capturing moving images, still images are taken at a constant time interval. Although the physical quantity "time" has an essentially continuous analog distribution, images are picked up using timings and lengths determined by imaging systems. This is the digitization of the time coordinate. The output of pixel r_k, color c_l, and time t_f^* is $S(r_k, c_l, t_f)$, and a set of S makes up one moving picture. As a result of the formation of built-in six-dimensional coordinate points, the information that is to be captured, which is essentially seven dimensional, is compressed to one dimension of light intensity information only. It is also a significant data compression.

As was discussed, on one hand, the information that constructs optical images is a set of continuous analog quantities (intensity, space, wavelength, and time), as shown Figure 1.9a. On the other hand, an image signal captured by image sensors is only the light intensity information arriving at the built-in coordinate points of space, wavelength, and time, as shown in Figure 1.9b. This is how image sensors work. The entity of the built-in coordinate points is each pixel covered with a one-color filter having a corresponding spectral distribution during the exposure period, as shown in the graphics in Figure 1.9b. A pixel output Sr_{mn} at address r_{mn} is expressed by

$$Sr_{mn} = \iiint\limits_{\Delta x \Delta y, \Delta \lambda, \Delta t} i(r,\lambda,t)\, f(r,\lambda)\, A(r,\lambda)\, dr d\lambda dt \tag{1.1}$$

where i, f, and A are the light intensity distribution of the optical image, the spectral distribution of the color filter, and the spectral sensitivity distribution of the image sensor, respectively. By integrating at the ranges of $\Delta x \Delta y$ in space, $\Delta \lambda$ in the wavelength coordinate and Δt in time, the light intensity signal is sampled. Therefore, the quality of the information on space, color, and time is decided by the sampling frequency and the width of the built-in coordinate point in (r, c, t) space. That is, it is determined in the wake of the system design. What remains is the accuracy and the dynamic range of the light intensity signal S.

The general coordinate points of space, wavelength, and time digitized in (r, c, t) space are shown in Figure 1.10a. The position coordinate r, color coordinate c, and time coordinate t are the pixel at that position, the color of the filter at the pixel, and the frame at the number of the exposure in chronological order, respectively. The number of the coordinate point of the space coordinate r, the color coordinate, and time coordinate is the same as the

---

* As will be discussed in Chapter 4, a single exposure time in moving pictures is called a frame.

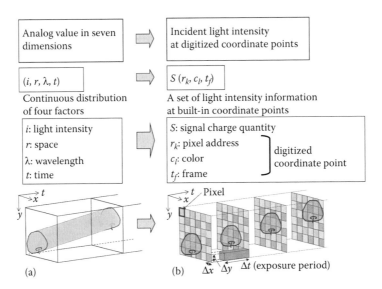

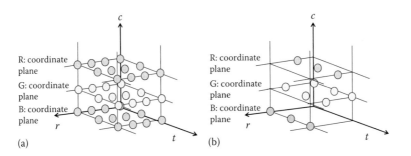

FIGURE 1.9
Comparison of optical image distribution with the image signal captured by image sensors: (a) optical image;
(b) image sensor signal.

FIGURE 1.10
Digitized coordinate points: (a) digitized general coordinate points in a color system with an RGB color filter;
(b) digitized coordinate points in a single-sensor color system with an RGB color filter.

pixel number in the image sensors, three in the case of the three primary colors system
using RGB and the number of frames, respectively. Since the color signal at the coordinate
point r_k contains only one color information of three in a single-sensor color imaging system, only one-third of the information is captured in (r, c, t) space, as shown in Figure 1.10b.
As the three color signals of RGB at each pixel are necessary to reproduce a color image,
the two color signals missing in the system are estimated by signal processing such as
demosaicking (color interpolation) using the correlation with the measured signal, as will
be discussed in Section 8.2.3.

While the signal of almost all image sensors is light intensity information at digitalized coordinate points, as shown in Section 1.2.3 by $S(r_k, c_l, t_f)$, different types of digitized coordinate points are possible. For example, as will be seen in Sections 5.3.3.2.3
and 7.3.2, there are sensors whose signal is time T, at which the change of light intensity
information reaches a predetermined amount as $T(f(i), r_k)$, where $f(i)$ is a function of light
intensity i.

1.3 Functional Elements of Image Sensors

Section 1.2 showed that the output of an image sensor is the intensity information of the incident light to each coordinate point. What functions must image sensors possess? First, they must have the function to measure the photon number, that is, the light intensity that comes to each coordinate point. Although, if it were possible,* an ideal method would be to count the photon numbers directly one by one; this is not realistic using the current technology. However, it can be achieved by using the following two steps. Initially, photons arriving at each coordinate point are absorbed to create signal charges (photoelectric conversion) by image sensors and the generated signal charges are integrated and stored (charge storage). In the second step, the stored charge quantity is measured and converted to electric signals (charge quantity measurement).

Secondary sensors must have the function to identify the signals from which coordinate point they emanate. Thus, image sensors must possess three functional elements: (1) a sensor part that generates and stores the signal charges; (2) a scanning part that identifies or addresses the coordinate point of the pixel of the signal; and (3) a signal charge quantity measuring part that measures the signal charge quantity and converts it to electric signals such as voltage, current, frequency, and pulse width.

Name of elements	1. Sensor part	2. Scanning part	3. Charge quantity measuring part
Function	Generation and storage of signal charges according to light intensity (light intensity information)	Addressing (identification of pixel address) (space information)	Measurement of signal charge quantity and conversion to electrical signals
Means	• Photoelectric conversion	• Charge transfer • X-Y addressing by shift register or decoder	• Charge–voltage conversion • Current–voltage conversion • Charge–frequency conversion • Charge–pulse width conversion

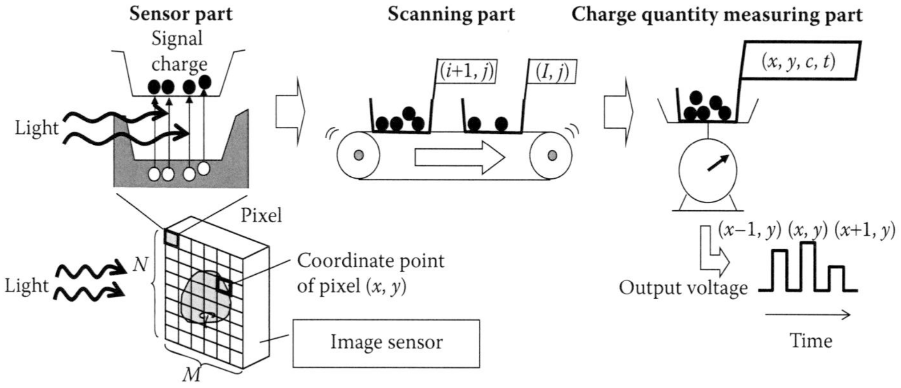

FIGURE 1.11
Functional elements of image sensors.

* It is possible to count the photon number directly only in the case of very low light intensity, when each photon comes with a sufficient time interval.

These elements are listed with their functions and material means in the table in Figure 1.11. The type of image sensor depends on the selection of the functional elements and their combination. The graphics in Figure 1.11 conceptually show the process by which the quantity information of the signal charges photoelectrically generated by the incident light to each pixel of an image sensor having $M \times N$ pixel number is transmitted with the address information and converted to electrical signals.

References

1. T. Kuroda, The 4 dimensions of noise, in *ISSCC 2007 Imaging Forum: Noise in Imaging Systems*, pp. 1–33, February 11–15, San Francisco, CA, 2007.
2. V. Gruev, A. Ortu, N. Lazarus, J. Spiegel, N. Engheta, Fabrication of a dual-tier thin film micro-polarization array, *Optics Express*, 15(8), 4994–5007, 2007.
3. T. Spirig, P. Seitz, O. Vietze, F. Heitger, The lock-in CCD-two-dimensional synchronous detection of light, *IEEE Journal of Quantum Electronics*, 31(9), 1705–1708, 1995.
4. H. Yabe, M. Ikeda, CMOS image sensor for 3-D range map acquisition using time encoded 2-D structured pattern, in *Proceedings of the 2011 International Image Sensor Workshop (IISW)*, p. 25, June 8–11, Hokkaido, Japan, 2011.
5. P. Seitz, Quantum-noise limited distance resolution of optical range imaging techniques, *IEEE Transactions on Circuits and Systems—I: Regular Papers*, 55(8), 2368–2377, 2008.
6. A. Wang, P.R. Gill, A. Molnar, An angle-sensitive CMOS imager for single-sensor 3D photography, in *Proceedings of the IEEE International Solid-State Circuits Conference Digest of Technical Papers (ISSCC)*, pp. 412–414, February 20–24, San Francisco, CA, 2011.
7. B.E. Bayer, Color imaging array, U.S. Patent 3,971,065, filed March 5, 1975, and issued July 20, 1976.

2

Device Elements and Circuits for Image Sensors

In this chapter, components of device elements commonly used in image sensors, silicon as a material for sensor parts, and circuit components are described.

2.1 Device Element Components

The term *components of device elements* here means the parts that form active semiconductor devices and are also components of large-scale integration (LSI) circuits. They are commonly used components, especially in image sensors. We begin with an explanation of the band structure of materials.

2.1.1 Foundation of Silicon Device Physics

Conceptual diagrams of the energy band structure of materials are shown in Figure 2.1.

According to the band theory of solid-state physics, electrons in crystals are arranged in energy bands named allowed bands, separated by the regions in energy space in which no electrons are allowed to exist. The energy width of this forbidden region is called the bandgap or energy gap. The states in allowed bands are occupied from the lowest to higher ones by electrons. The highest energy level of the occupying electron is the concept of Fermi level.*

Electric current means the movement of electrons (charged particles) in real space. In the allowed bands of some materials, states are partly filled, as shown in Figure 2.1a. As the upper part of the band is energetically in continuance and has some space, acceleration of the electric field can raise the energy a little, and electrons can rise to the upper states. Thus, the electron distribution can move as a whole in energy space. That is, the electric current can flow. On the contrary, in materials whose bands are filled or empty, as shown in Figure 2.1b, the acceleration described above cannot occur. Since the energetic width of the bandgaps prevents electrons from rising to upper empty bands, even if they try to get a little energy from the electric field, there is no possible state to transit to within the accelerated energy region. Therefore, electric current cannot flow in this case. Thus, Figure 2.1a shows a conductor such as metal, while Figure 2.1b shows an insulator or semiconductor at absolute zero. Insulators and semiconductors are categorized in the same group in this respect, although the bandgap of an insulator is very wide, while that of a semiconductor is narrow.

A band in which all states are filled with electrons is the valence band and one in which all states are unfilled is the conductive band.†

---

* This is a conceptual expression and it is expressed more accurately as the parameter in Fermi distribution function.

† The energy value of the vertical axis is significant in Figure 2.1, while the horizontal one has no meaning.

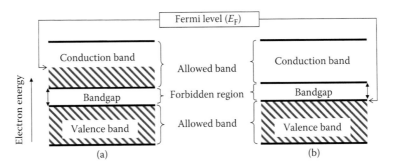

FIGURE 2.1
Schematic energy bands and electron occupancy of allowed band: (a) conductor (metal); (b) semiconductor at absolute zero and insulator.

The difference between semiconductors and insulators lies in the width of the bandgap. As the bandgap of semiconductors is small, the density of electrons excited* from the valence band to the conduction band by thermal energy of 26 meV at room temperature 300 K is substantial.[†] On the contrary, the bandgap of insulators is wide, so the density of electrons excited to the conductive band is negligible. Thus, electrical conduction is low and electrical resistance is very high.

In metals, electron density, which contributes to electric current, is very high and there is no external means to control it. On the contrary, it can be easily controlled in semiconductors. This is the important source of the function of semiconductor devices.

The material of most real semiconductor LSI circuits is silicon. In silicon devices, electrons that are excited from the valence band to the conductive band, as mentioned above, are infrequently used. When silicon is used as a starting material for LSI circuits in manufacturing, its purity is raised up to 99.999999999% (eleven nines) and a different atom is added afresh.[‡] However it is added intentionally, it is customarily called an impurity.

As silicon belongs to the IV family of the periodic table, it has four valence electrons, each of which forms a covalent bond with its nearest neighbor in the silicon crystal, as shown in Figure 2.2.

As phosphorus (P) and arsenic (As) belong to the V family, they have five-valence electrons. If one silicon atom in the crystal is displaced by one phosphorus or arsenic atom, four electrons on the phosphorus or arsenic atom form covalent bonds, similar to silicon, as shown in Figure 2.3. The fifth electron is bound by a positively charged phosphorus or arsenic ion (P^+, As^+) with a binding energy of about 45 meV. However, this electron has a kinetic energy of 26 meV at room temperature; it leaves the bind easily and moves. Thus, the fifth electron contributes to conduction. The phosphorus and arsenic atoms are called donors because they donate an electron to the conduction band by ionization.

The case of applying atoms of the V family has been explained above. But what about the III family? Boron belongs to the III family and has only three valence electrons. This means one bonding electron is missing when a silicon atom is replaced by a boron atom in a silicon crystal, as Figure 2.4 shows.

---

* Thermal energy as kinetic energy is converted to potential energy.
† At absolute zero, semiconductors do not exist but insulators do.
‡ This is called doping, and non-doped silicon is called intrinsic silicon.

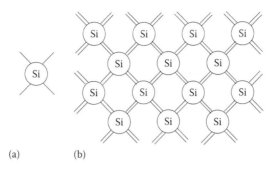

FIGURE 2.2
(a) Schematic diagram of silicon atom; (b) diagram of silicon crystal.

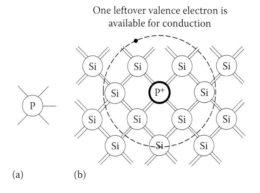

FIGURE 2.3
Schematic diagram of phosphorus-doped silicon crystal: (a) schematic diagram of phosphorus atom; (b) diagram of phosphorus atom in silicon crystal.

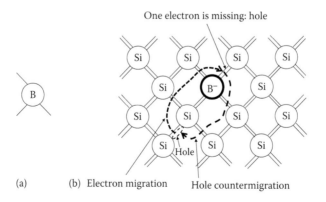

FIGURE 2.4
Schematic diagram of boron-doped silicon: (a) boron atom; (b) boron atom in silicon crystal.

Boron can complete its bonds only by taking an electron from a Si–Si bond, as shown in Figure 2.4b, leaving behind a hole in the silicon valence band. It is called a hole because it is a state that an electron should occupy but is missing from and is positively charged due to the absence of an electron. When another electron moves and fills it in, it can be seen that the hole moves in the counterdirection, which is regarded as a movement of the hole.

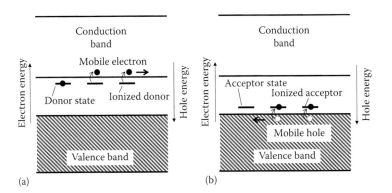

FIGURE 2.5
Schematic energy diagram: (a) *n*-type silicon, (b) *p*-type silicon. ●, electron; o, hole.

A hole is bound by a negatively charged boron ion (B⁻) with a binding energy of 45 meV. It can be easily freed from the bind by thermal energy and moves in the same way as electrons of phosphorus and the arsenic atom. The boron atom is called an acceptor because it accepts an electron from the valence band through ionization.

While Figures 2.3 and 2.4 show spatial appearance, Figure 2.5 shows the distributions of electron and hole in energy space. Figure 2.5a shows diagrammatically that electrons bound at donor states formed by phosphorus or arsenic leave the bind by thermal excitation to the conduction band to be mobile electrons. Depending on the necessity, the density of the doped impurity is around 10^{14}–10^{18} cm³, that is, about one impurity atom to 10^5–10^9 silicon atoms. Although the doped impurity atom is electrically neutral, the binding energy of the electron to the ionized donor is of such a level that the electron becomes free from the bound state by thermal excitation at room temperature. Ionized donor impurities* are positively charged by the release of electrons and the same numbers of mobile electrons are excited to the conduction band.

As with phosphorus and arsenic, boron is fixed in silicon crystals, as shown in Figure 2.4, and becomes a negatively charged ionized acceptor by accepting an electron. It generates the same numbers of holes in the valence band, as shown in Figure 2.5b.

Thus, the semiconductor whose mobile charge is initiated by electrons is called an *n*-type semiconductor because of its negative charge polarity, and the doped atoms of phosphorus arsenic are called an *n*-type impurity. When the hole of the positive charge is the initiator, this is called a *p*-type semiconductor, and boron as the element is called a *p*-type impurity.

The electron and hole are called carriers as they carry the charge that flows to create an electric current. The directions of carrier flow and electric current are the opposite for electrons because of their negative charge but the same for holes. While electrons in *n*-type semiconductors and holes in *p*-type semiconductors are called majority carriers, holes in *n*-type semiconductors and electrons in *p*-type semiconductors are called minority carriers.

2.1.2 *pn*-Junction

As Figure 2.6 shows, the structure of a *pn*-junction is literally the connection of a *p*-type semiconductor area with an *n*-type semiconductor area. In practical fabrication, *n*-type impurity atoms whose concentration is more than one digit higher than that of the *p*-type ones are

---

* They are fixed spatially in the silicon crystal, as shown in Figure 2.3.

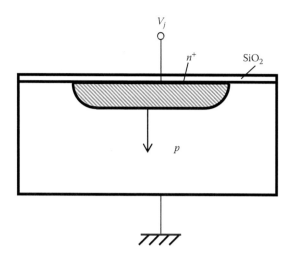

FIGURE 2.6
Cross-sectional view of *pn*-junction.

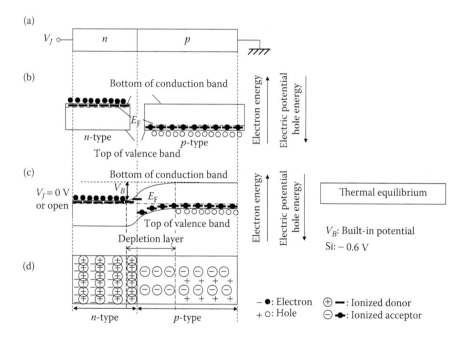

FIGURE 2.7
Potential distribution model of *pn*-junction in thermal equilibrium: (a) model structure; (b) band diagram in separated situation; (c) band diagram of *pn*-junction; (d) spatial distribution in *pn*-junction.

doped into a part of the *p*-type semiconductor area. In contrast, doping high-concentration *p*-type atoms into an *n*-type semiconductor area is also possible.

Although both *n*-type and *p*-type impurity atoms are distributed in the doped area, the type having the higher concentration determines the polarity of conduction of the area, because electrons occupy the states from the lowest to highest in energy space.

Using the model in Figure 2.7a, which shows *n*-type and *p*-type regions are directly connected in real space, we show how energy distribution or potential distribution is formed in the *pn*-junction. The *p*-type region is connected to ground level and voltage source V_J is applied to the *n*-type region. Figure 2.7b shows that the *n*-type region and *p*-type region exist separately in energy space. Ionized donors and electrons as carriers are equally distributed in the *n*-type region and ionized acceptors and holes are equally distributed in the *p*-type region. All areas are electrically neutral. As the Fermi level is situated in the vicinity of the maximum energy of the electron distribution, it is near the lower level of the conductive band in the *n*-type region and near the upper level of the valence band in the *p*-type region. Then, if the *n*-type region and *p*-type region are connected, electrons in the vicinity of the junction in the *n*-type region and holes in the vicinity of the junction in the *p*-type region diffuse toward the opposite areas, according to the diffusion voltage caused by the concentration difference between both sides. Diffusion of electrons and holes continues till the electric field formed by the remaining ionized impurities increases to balance out the diffusion voltage. Consequently, as Figure 2.7c shows by the potential distribution, ionized donors remain in the *n*-type region near the junction and ionized acceptors remain in the *p*-type region, that is, only spatially fixed charges remain, and no carrier exists continuously in this area. Therefore, this area is called the depletion layer.

The appearance of diffusion stops and an equilibrium is reached. Figure 2.7d shows this situation by the spatial distribution, while Figure 2.7c shows it in energy space. In the depletion layer, the total number of positive charges in the *n*-type region and negative charges in the *p*-type region are equal. Outside the depletion layer in both the *n*-type region and *p*-type region, mobile and fixed charges exist equally so that they are electrically neutral. The difference in potential between the *n*-type region and *p*-type layer is called the built-in potential and is decided by the impurity concentrations of both sides and temperature. It is about 0.6 V in silicon at room temperature.

Bias condition is discussed next. While the *p*-type region in the *pn*-junction is grounded, positive voltage V_J is applied to the *n*-type region, as shown in Figure 2.8a as potential and charge distribution and in Figure 2.8b as distribution in space.

In the *n*-type area, electrons drain and decrease until the voltage increases to V_J. It is also believed that the decrease of negatively charged electrons creates a positive voltage in the *n*-type region. As a result, the width of the depletion layer increases on both sides of the junction, and no current flows through the *pn*-junction. When positive voltage is applied to the *n*-type region, the state of the *pn*-junction is called reverse-biased condition. As will be mentioned later, as photodiodes in image sensors, *pn*-junctions are used by being made electrically separate and floating in a reverse-biased condition, to integrate signal charges generated by incident light. At the instant of electrical separation, it is in nonequilibrium, and it makes use of a phenomenon returning to an equilibrium situation.

In contrast to the above case, when negative voltage is applied to the *n*-type region, V_J as applied voltage decreases from 0 V and when the absolute value becomes the same as the built-in potential, that is, $V_J = -|\phi_B|$, the band energies of the *n*-type and *p*-type regions become equal. At the point when V_J becomes large in the negative direction, electron energy in the *n*-type region becomes higher than in the *p*-type region, as Figure 2.8c shows; then the flow of electrons starts from the *n*-type region to *p*-type region. At the same time, holes are now in the same situation and flow from the *p*-region to *n*-region. Thus, electric current flows in this bias condition. This bias state is called forward bias.

Although electric current flows through the *pn*-junction under forward bias, it does not occur under reverse bias, that is, *pn*-junctions have a rectifying action. As discussed, the names of reverse and forward biases come from this function.

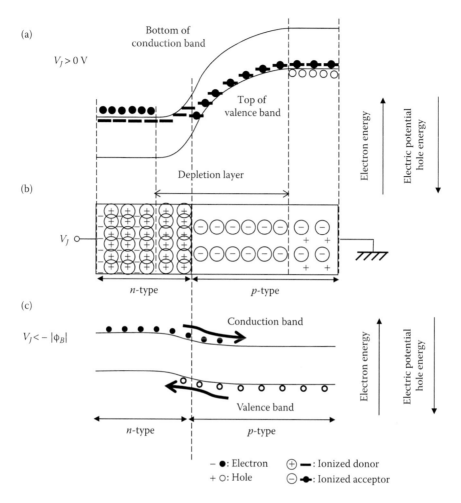

FIGURE 2.8
Potential distribution model of biased *pn*-junction: (a) potential and charge distribution of reverse-biased *pn*-junction; (b) spatial distribution of charge of reverse-biased *pn*-junction; (c) carrier flow through forward-biased *pn*-junction.

2.1.3 MOS Structure

As a typical example of a metal-oxide semiconductor (MOS) structure, a cross-sectional view of a gate electrode–silicon dioxide film (SiO_2)–*p*-type silicon structure is shown in Figure 2.9. A conductive gate electrode is formed on silicon via silicon dioxide film.

Let us consider the behavior from the viewpoints of spatial and energetic distributions using Figure 2.10. Figure 2.10a and b show spatial and energetic distribution of charges, respectively, at the state that the gate voltage V_G is 0 V. The same voltage is applied to both the *p*-type Si and the gate electrode. It is supposed that there is no potential difference between the silicon and the gate electrode in this condition. This is called a flat-band condition. In real devices, the gate voltage at a flat-band condition generally deviates from 0 V because of the difference of concentration of impurities between semiconductor and gate electrode materials, and also because of the existence of ions in the silicon oxide film. However, these are ignored to simplify the arguments in this book.

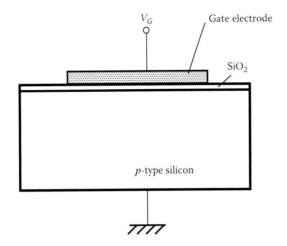

FIGURE 2.9
Cross-sectional view of MOS structure with gate electrode–silicon dioxide film (SiO₂) *p*-type silicon structure.

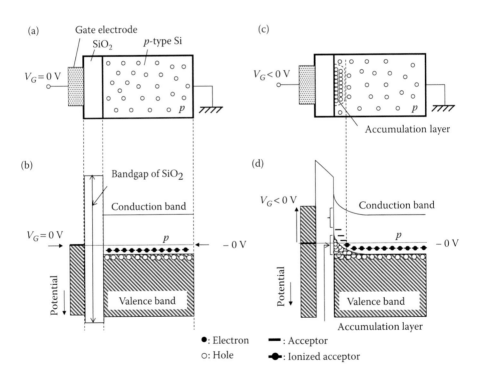

FIGURE 2.10
Spatial and potential distribution model of MOS structure at various gate biases: (a) spatial distribution (flat-band condition); (b) potential distribution (flat-band condition); (c) spatial distribution (surface accumulation layer); (d) potential distribution (surface accumulation layer); (e) spatial distribution (shallow depletion); (f) potential distribution (shallow depletion); (g) spatial distribution (deep depletion, without electron supplier to surface); (h) potential distribution (deep depletion); (i) spatial distribution (surface inversion layer, with electron supplier to surface); (j) potential distribution (surface inversion layer).

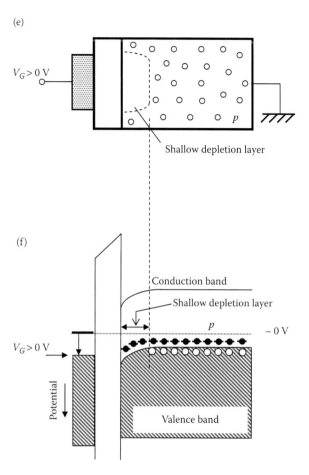

FIGURE 2.10 **(Continued)**

First, when negative voltage is applied to the gate electrode, the holes gather in the silicon surface because of their positive charge, as Figure 2.10c shows by spatial distribution. As shown by energy distribution in Figure 2.10d, the holes in the p-type semiconductor are attracted to the lower energy state formed by the negative potential of the gate electrode and accumulated in the silicon surface. This hole layer is called the surface accumulation layer. The electron layer accumulated in the silicon surface by the application of a positive voltage to the gate electrode in the n-type silicon MOS state is also called the surface accumulation layer.

Next, as Figure 2.10e shows by spatial distribution, the application of a low positive voltage, less than about 1 V, to the gate electrode causes holes to take off from the surface and only the negatively charged ionized acceptors remain, forming a depletion layer. In Figure 2.10f, this is shown as an energy distribution: holes escape from the higher potential surface area caused by the positive potential of the gate electrode.

The higher positive voltage application to the gate electrode indicated in Figure 2.10g shows that the depletion layer extends deeper inside the semiconductor and the width (depth) enlarges as spatial distribution, as also shown in energy space in Figure 2.10h. The rough indication of deep depletion is thought to be a gate voltage higher than about 1.1 V, which corresponds to the silicon bandgap at room temperature. In this situation, if

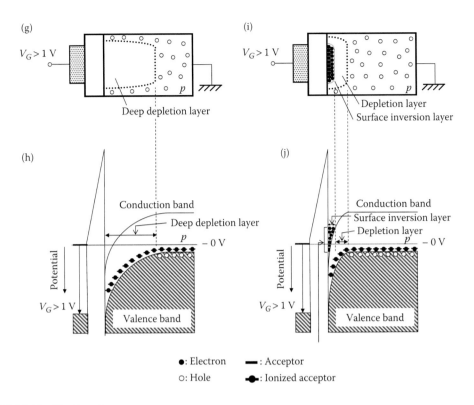

FIGURE 2.10   (Continued)

electrons are supplied into the conduction band by some means, electrons gather at the hollow of potential at the surface, as Figure 2.10i and j show. The width of the depletion layer shrinks so that the total area density of negatively ionized acceptors and electrons is maintained. Since electrons have the opposite polarity to *p*-type semiconductors, this electron layer is called the surface inversion layer.

As will be described, when the means to supply electrons is light in the above situation, it plays a sensing role and is named a photogate sensor.

In contrast, the hole layer collected on the silicon surface by the application of a negative voltage to the gate electrode in an *n*-type silicon MOS state is also called an inversion layer.

The MOS field-effect transistor (MOSFET) is a device to control carrier density in the surface inversion layer by applying voltage to the gate electrode. Figure 2.11 shows that, adjacent to both sides of the channel under the gate electrode in the MOS structure, the highly concentrated *n*-type regions are formed as a source for electrons and also as a drain to accept electrons. The voltage applied to a gate electrode controls the density of current. As the electrons carry electric current in a MOSFET of *p*-type substrate, it is called an *n*-channel or *n*-type MOSFET. On the other hand, a MOSFET composed of *n*-type substrate, source, drain of *p*-type region with negative voltage, and holes whose density at the channel is controled by negative voltage applied to a gate electrode is called a *p*-channel or *p*-type MOSFET.

2.1.4 Buried MOS Structure

Following the ordinal MOS structure, the buried MOS structure, which is used in buried-channel charged-couple devices (CCDs) and buried MOSFETs, is described.

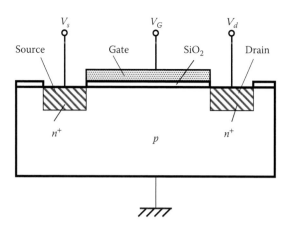

FIGURE 2.11
Cross-sectional view of MOSFET.

In contrast with the ordinal MOS structure, in the buried MOS structure the *n*-type layer is formed in the channel area under the gate electrode, having antipolarity and lower impurity concentration compared to the *p*-type substrate, as shown in Figure 2.12.

A *pn*-junction is formed with low impurity concentration of *n*-type layer at the silicon surface.

It is considered how this *n*-type area plays a role in Figure 2.13, where (a) indicates charge distribution in real space, while (b) indicates charge distribution in energy space. Ground-level voltage is applied to both the *p*-type substrate and gate electrode. The system is in thermal equilibrium, and a built-in potential with a depletion layer exists between both sides of the *pn*-junction. In other areas, ionized impurity, which is spatially fixed, and an antipolarity carrier are distributed and are electrically neutral, as shown in the figure.

Here, if electrons in the *n*-type layer are pulled out, for example in such a way as to set up a higher-concentration *n*-type drain adjacent to the *n*-type layer and apply a higher

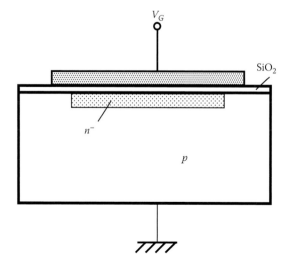

FIGURE 2.12
Cross-sectional view of buried MOS structure.

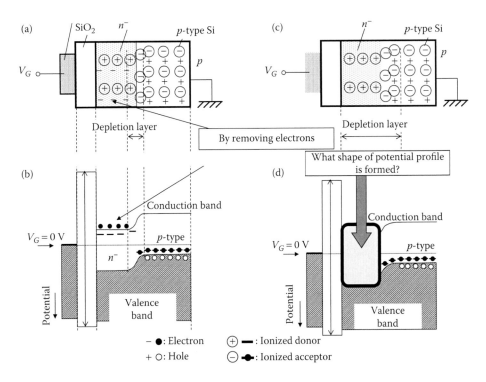

FIGURE 2.13
Charge and potential distribution of buried MOS structure: (a) spatial distribution (not depleted); (b) potential distribution (not depleted); (c) spatial distribution (fully depleted); (d) potential distribution (fully depleted).

positive voltage, the *n*-type area will be completely depleted and only positively charged ionized donors spatially fixed will be distributed, as indicated in Figure 2.13c. How is the potential distribution in the *n*-type layer indicated in Figure 2.13d?

Here, a very simplified way to consider this situation is introduced, as shown in Figure 2.14a.[1]

1. The spatial distribution of positive charges (ionized donors) is uniform.
2. Both sides of the *n*-type layer are grounded, that is, the potential profile is bilaterally symmetric, including the boundary conditions.

The electric potential ϕ of this area is expressed by the Poisson equation as follows:

$$\frac{d^2\phi}{dx^2} = \frac{-\rho}{\kappa\varepsilon_0} \ (\text{constant})$$

(2.1)

$$\rho = eN_D$$

where *x* is depth
ρ = charge density
κ = relative permittivity of silicon

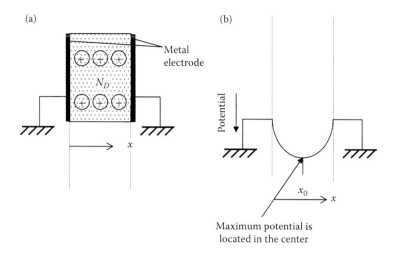

FIGURE 2.14
Simplified model of potential profile of buried MOS structure: (a) spatial distribution; (b) potential distribution.

ε_0 = vacuum permittivity
e = elementary charge
N_D = donor density

Thus, as a quadratic differential is a constant value, the potential distribution ϕ is expressed as follows:

$$\phi = \frac{-eN_D}{2\kappa\varepsilon_0}(x - x_0)^2 + c \tag{2.2}$$

Therefore, as Figure 2.14b shows, the potential profile is expressed by a quadratic function of x, whose curvature is in proportion to donor density and convex downward, and the maximum value is c at the coordinate point $x=x_0$. This means the potential maximum is located in the center. Downward is defined as potentially positive here as usual. The boundary condition in a real device is not bilaterally symmetric, which means that the selected part of a quadratic function curve only shifts to either side according to the boundary condition and is really a part of the quadratic function curve that Equation 2.2 shows. The nonuniformity of the impurity density in a real device only causes partial distortion from a perfect quadratic curve.

As a result, the potential profile in the n-type layer, as has been considered in Figure 2.13d, is understood to be a convex downward quadratic curve, as shown in Figure 2.15.

It is important that the maximum point of the potential is located inside the silicon, separate both spatially and energetically from the silicon–oxide interface so that the electrons can exist and pass without touching the interface. In practice, this enables avoidance of the influence of the interface state existing in the interface, which will be described as a basic structure of the buried CCD in Section 5.1.1.

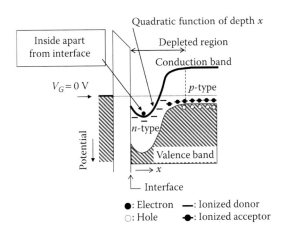

FIGURE 2.15
Schematic potential profile of buried MOS structure.

2.1.5 Photogate

Photosensitive element components are now described. The characteristics of silicon as a photosensitive material will be explained in Section 2.2. It is assumed that the signal charge is an electron unless otherwise noted.

Photogate sensors are MOS structures, as shown in Figure 2.16a. The application of positive voltage to the gate electrode forms a depletion layer with the maximum potential at the surface, as shown in Figure 2.16b. The light transmitted through the gate electrode is absorbed in silicon and generates a pair of an electron and a hole. The electric field in a depletion layer separates them spatially. Electron–hole pairs created in deeper positions in substrate where there is no electric field tend to vanish by recombination because there is no force to separate them, shown as generation and recombination in Figure 2.16b. The electrons are collected in the potential well at the surface, and the holes move to the substrate side to be discharged finally outside the device. Here the gate electrode must allow light to pass through, so metals cannot be used as the

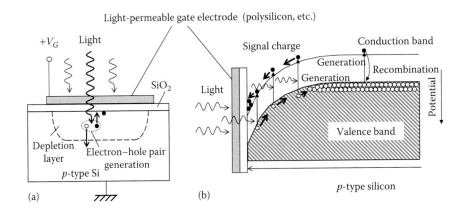

FIGURE 2.16
Photogate: (a) cross-sectional view; (b) operation in energy space.

material for the electrode. There are a few examples of transparent electrodes such as indium tin oxide (ITO), but polysilicon, which is heavily used in the processes of silicon LSI circuit production, is the most commonly used. The light-absorbing characteristic of polysilicon is basically the same as that of silicon. Section 2.2 discusses how the absorption coefficient at longer wavelengths is lower. However, it tends to be higher for shorter wavelengths, so that the amount of blue light that reaches silicon is reduced and the sensitivity tends to be low.

2.1.6 Photodiode

The structure of the photodiode is the same as that of the *pn*-junction, as Figure 2.17 shows, and it does not have a gate electrode, which attenuates the intensity, especially of blue light. In operation, positive voltage is applied to the *n*-type area to create a reverse condition, and the reverse-biased *pn*-junction is isolated electrically and left in a floating state, as mentioned in Section 2.1.2. The *n*-type region is not completely depleted in an ordinary photodiode. The energy of light that silicon absorbs excites an electron from the valence band to the conduction band and a hole is left in the valence band. Electrons flow into the highly potential *n*-type region and are stored, and holes flow toward the substrate and are discharged. The behavior in silicon is the same as in the photogate and photodiode; while only the method of biasing is different.

2.1.7 Buried Photodiode/Pinned Photodiode

A buried or pinned photodiode is used in most of sensors for high image quality applications. The difference from an ordinary photodiode is that it has a *pnp* structure created by forming a high concentration of p^+ layer at the surface, the *n*-type region is completely depleted, and the potential is pinned in the depleted region, as shown in Figure 2.18.

The potential profile in silicon is the same as with a buried MOS structure, as discussed in Section 2.1.4. It will be explained in Section 5.1.2.3 that the existence of a high-concentration hole layer facing the interface greatly suppresses dark output (dark current) noise generation by the electrons, which are thermally excited to the conduction band through the interface state on the silicon oxide interface. The absorption of light and the integration of signal charges in silicon are the same as in an ordinary photodiode. The names buried and pinned photodiode originate from the structure and the function, respectively.

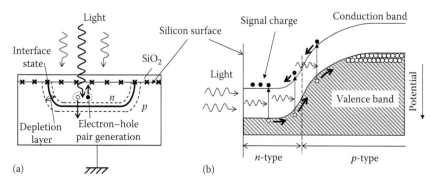

FIGURE 2.17
Photodiode: (a) cross-sectional view; (b) operation in energy space.

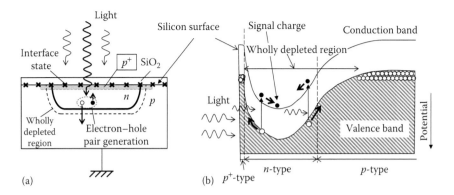

FIGURE 2.18
Buried/pinned photodiode: (a) cross-sectional view; (b) operation in energy space.

2.2 Silicon as a Photosensitive Material

Silicon is a key material of LSI circuits, and image sensors are LSI circuits. Also, silicon absorbs light of ultraviolet, visible, and near-infrared wavelength regions to generate charge. Therefore it is natural that silicon has come to be used in photosensitive parts in image sensors. However, this does not mean that silicon is the best photosensitive material for all image sensors.

As Figure 2.19 shows, the luminous flux density ϕ travels through silicon being absorbed. The luminous flux density reduces the amount of $\Delta\phi$ while advancing the distance of Δx because of absorption. Expressing absorption coefficient α, the relationship is shown as follows:

$$\Delta\phi = -\alpha\phi\Delta x \tag{2.3}$$

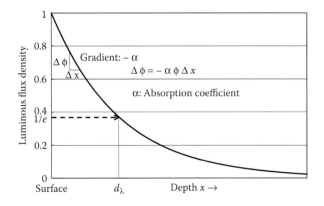

FIGURE 2.19
Depth dependence of luminous flux density.

This leads to the following equations:

$$\frac{\Delta\phi}{\Delta x} = -\alpha\phi, \quad \frac{\mathrm{d}\phi}{\mathrm{d}x} = -\alpha\phi$$

$$\phi = \phi_0\, e^{-\alpha x} = \phi_0\, e^{-x/d_\lambda} \quad \left(d_\lambda = 1/\alpha\right) \tag{2.4}$$

where ϕ_0 denotes the luminous flux density at the silicon surface just before absorption, the inverse of absorption coefficient α; and d_λ is penetration depth, indicating the distance through which luminous flux density attenuates to $1/e$ (e is the base of natural logarithm) by absorption. This is an indication of the length required to absorb light to a certain degree.

Figure 2.20 indicates the wavelength dependence of the absorption coefficient and penetration depth of silicon. It can be seen that it needs a longer distance to adequately absorb longer-wavelength light. Compared with red light (around 640 nm), where d_λ is about 3–4 μm, blue light (around 440 nm) is as short as 0.3 μm. This means that the depth required for a photodiode to absorb light efficiently differs greatly, according to the wavelength. For the use of color imaging, the depth of the photodiode is ordinarily set to red light, which needs the longest distance lest the sensitivity should be reduced.

2.2.1 *np* Photodiode on *p*-Type Substrate

The measured example of spectral response of an *np* photodiode that is formed on a *p*-type substrate, whose cross-sectional diagram is shown in Figure 2.21a, is indicated

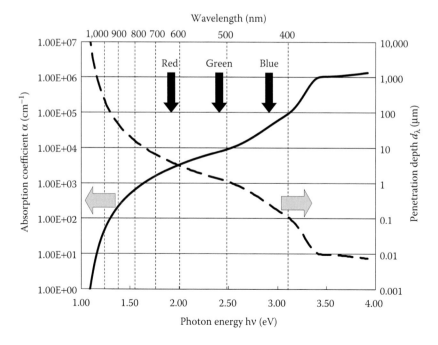

FIGURE 2.20
Wavelength dependence of absorption coefficient and penetration depth of silicon.

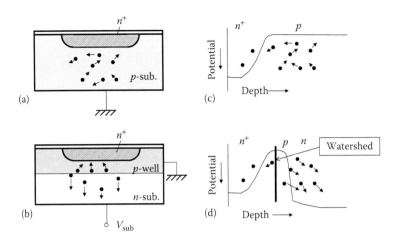

FIGURE 2.21
Comparison of *np* and *npn* photodiodes in structure and potential profile: (a) *np* photodiode on *p*-type substrate; (b) *npn* photodiode in *p*-well on *n*-type substrate; (c) potential profile of *np* photodiode; (d) potential profile of *npn* photodiode.

by the solid line of the *np* photodiode (*p*-type substrate) in Figure 2.22. The absorption is observed from a little less than 400 nm to the interband absorption edge of silicon, including the visible region (380–780 nm) and near infrared. Especially strong absorption is seen in the near-infrared region, whose wavelength is longer than that of the visible region. This means silicon is a superior photosensitive material in the near-infrared region. However, as for the sensors of color cameras based on the visible range, high sensitivity in the near-infrared region is rather undesirable because it disrupts color balance.

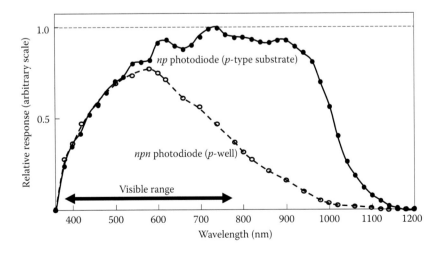

FIGURE 2.22
Measured examples of spectral response of *np* and *npn* photodiodes.

2.2.2 *npn* Photodiode on *p*-Well

A photodiode that is formed in a *p*-well on an *n*-type substrate lowers the unnecessary sensitivity of near-infrared light, as mentioned above. Figure 2.21b indicates an *npn* photodiode[2] that is formed in a *p*-well on an *n*-type substrate. The *p*-well is grounded and positive voltage is applied to the *n*-type substrate. Figure 2.21c shows the potential distribution of an *np* photodiode and Figure 2.21d shows that of an *npn* photodiode. In the case of the *np* photodiode, it is clear that the signal charges that are generated in the deeper region of the substrate seem to be collected and in a photodiode, this can contribute to the sensitivity. Meanwhile, in an *npn* photodiode, the charges that are generated in the area deeper than the least potential point of the *p*-type region (which is indicated by "Watershed" in the figure) cannot come up to the surface but are emitted to the substrate and discharged, so they do not contribute to the sensitivity. The spectral response of the *npn* photodiode is indicated by a dashed line in Figure 2.22. In the case of the *npn* photodiode, the near-infrared sensitivity, which penetrates into the depth of the substrate because of the low absorption coefficient, is shown to be greatly reduced.

2.3 Circuit Components

The circuit components that are commonly used in image sensors will be described in this section.

2.3.1 Floating Diffusion Amplifier

A floating diffusion amplifier (FDA)[3] is a component that measures the amount of electric charge, which is explained in Figure 2.23.

As Figure 2.23a shows, an FDA consists of a capacitor, a reset transistor that resets the potential of the capacitor to a power-supply voltage V_d, and an amplifier that receives the potential of the capacitor to produce voltage output. It operates as follows:

1. By applying a reset pulse to the gate electrode of the reset transistor to be on-state, capacitor C is connected to V_d, and by switching the reset of the transistor to off-state, the potential of capacitor C is reset to V_d, as Figure 2.23b shows.
2. The potential of capacitor C at this moment is received by the amplifier to produce an output.
3. By transferring signal charges with quantity Q to capacitor C from the above state, Q will make the potential of capacitor C shallower than V_d by a signal voltage of $Q/C = V_Q$ as Figure 2.23c shows.
4. The amplifier receives the potential of the capacitor at this state and produces an output.

This means that the potential difference between (2) and (4) is the signal voltage that is proportionate to the charge amount Q.

After completion of a measurement, a reset action is done again to accept the subsequent signal charges. From the principle of action, as Figure 2.23d shows, if the volume of

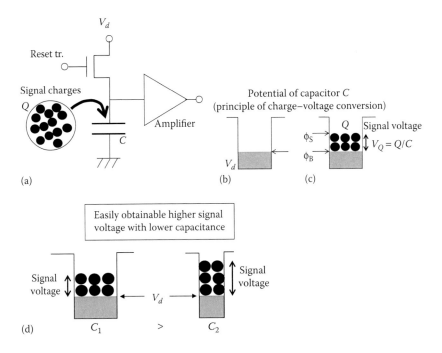

FIGURE 2.23
Working principle of FDA: (a) configuration; (b) electric potential of the capacitance just after reset operation; (c) electric potential of the capacitance after reception of signal charges; (d) capacity and signal voltage.

capacitor C becomes lower, the higher signal voltage is easily obtained and this is advantageous for the later signal processing from the viewpoint of signal-to-noise ratio.

On the other hand, if the capacitor volume is too low, the maximum charge amount in this capacitor is limited because of the saturation under the treatable voltage range. And if a high voltage range is adopted, the output will be out of the voltage range at a later stage. As these conditions sometimes limit the saturation characteristics, a balanced design is required. Since this capacitor becomes electrically isolated to be floating after being reset to sense the potential change by the introduction of the signal charges, it is called floating diffusion (FD).

In the circuit of Figure 2.23a, the amplifier that is commonly used to sense the potential of capacitor C is a source follower amplifier (SFA), which is discussed next.

2.3.2 Source Follower Amplifier

A source follower amplifier (SFA) is a so-called *buffer circuit*, which accepts the input signal potential at the gate input part with high input impedance of MOSFET and produces the same potential of output with low output impedance. Other circuits for the same kinds of purposes are emitter follower circuits, which use bipolar transistors, and voltage follower circuits, which use operational amplifiers. SFAs accept the input voltage V_{in} at the gate input part and produce the output voltage V_{out} at the source output part.

As Figure 2.24a shows, the drive transistor (MOSFET) with the voltage input part and load devices are connected in series in SFAs. Load devices include elements such as a constant current source, shown in Figure 2.24b, a resistor, shown in Figure 2.24c, and a load transistor, shown in Figure 2.24d. Each voltage–current characteristic of the drive

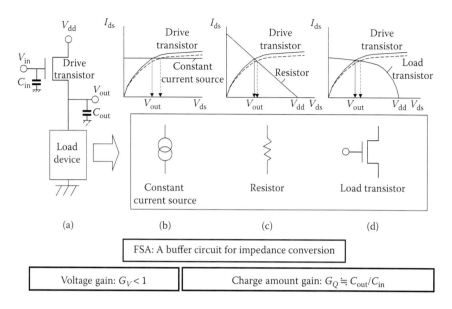

FIGURE 2.24
Source follower amplifier: (a) general configuration; (b) constant current source; (c) resistor; (d) load transistor.

transistor and load device is shown in Figure 2.24. As the currents through the drive transistor and load device are the same, the intersection point of the two voltage–current curves is the operating point. If the signal voltage of the input V_{in} in SFA changes, the operating point moves. The solid curved lines of the voltage–current characteristic of the drive transistor in Figure 2.24b–d are the characteristics before the input voltage of V_{in} changes, that is, a no-signal situation. The broken lines show those after the change, that is, the signal charge quantity information is added to input V_{in}. Therefore, the voltage difference between two intersection points on the voltage–current characteristic curve indicates the output voltage amplitude of SFAs. Regarding the output change in contrast with the input voltage change, the constant current source is the highest of the three. Although it has a good characteristic, the area of circuitry tends to be large, and therefore the load transistor is used in most cases because of characteristic—circuit size trade-offs.

An ideal characteristic is that the output voltage V_{out} is equal to the input voltage V_{in}, however, output voltage is obtained by subtracting the threshold voltage V_{th} of the drive transistor from the input voltage and multiplying by SFA voltage gain G_V (<1, around 0.7–0.9). This means the actual output voltage is lower than the input one, as expressed by the following equation:

$$V_{out} = (V_{in} - V_{th})G_V \tag{2.5}$$

Despite the name "amplifier," the voltage gain is less than unity, as mentioned above. Their respective load capacitors, C_{in} and C_{out}, are at the input part and the output part of the SFA. Under a very simplified condition, $G_V = 1$, $V_{th} = 0$, and $V_{in} = V_{out}$. Then the amount of charge accumulated in C_{out} is C_{out}/C_{in} times the charge amount in C_{in} because of the same voltages. Although the voltage gain of SFA is less than unity, the amount of charge is greatly multiplied and this is important to drive the later circuit.

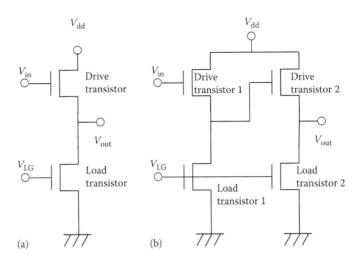

FIGURE 2.25
Configuration of SFA: (a) single-stage configuration; (b) dual-stage configuration.

According to the necessary frequency characteristic and the necessary ability to drive the following stage circuit, SFA is sometimes used with single-stage and other times with dual- or triple-stage cascade arrangements, as Figure 2.25 indicates.

2.3.3 Correlated Double Sampling Circuit

When signal charges Q are introduced in FD, the essential information is V_Q in Figure 2.23c, because it has the direct information about charge quantity Q. Therefore, the difference between the potentials of the capacitor C before and after reception of signal charge is $(\phi_S - \phi_B)$, as shown in Figure 2.23b and c. A circuit to get the difference between the two is a correlated double sampling (CDS) circuit. The reset levels in Figure 2.23b and c are the same, because the level is set by a single reset operation, that is, they are correlated to each other.

In actuality however, a small charge amount Q cannot drive a CDS circuit. So, after multiplication of the amount of signal charge Q by SFA, the CDS operation is achieved. Figure 2.26 shows that the output of the FDA, which uses an SFA as a buffer, is followed by a CDS circuit.

In this way, the difference of output of SFA between input ϕ_S and ϕ_B is expressed as follows, referring to Equation 2.5:

$$(\phi_S - V_{th})G_V - (\phi_B - V_{th})G_V = (\phi_S - V_B)G_V \tag{2.6}$$

that is, the input potential difference $(\phi_S - \phi_B)$ multiplied by SFA voltage gain G_V; and the threshold voltage of the drive transistor, V_{th}, has vanished. As will be discussed in Chapter 5, an SFA in a CCD has little influence on the value of V_{th} because only one SFA is commonly utilized. However, in cases when SFAs are arranged in each pixel like CMOS sensors, the variation of V_{th}, which can be as low as a few millivolts, will have a serious influence as a fixed-pattern noise on the image quality so that the disappearance of the V_{th} influence is very important.

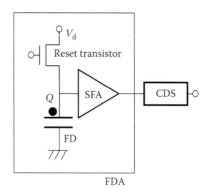

FIGURE 2.26
Block diagram of FDA and CDS circuit connection.

The principle construction of a CDS circuit[4] is shown in Figure 2.27a. The input follows the SFA output in an FDA and an alternating current (AC) coupling capacitor C_c transfers only the AC component to node A. A clamp transistor that connects node A and the clamp voltage V_{clamp} source is formed, and clamp pulse ϕ_{CL} is applied to the gate electrode. A sampling transistor to which sampling pulse ϕ_{SH} is applied at the gate electrode and capacitor C_{SH} is formed to sample and hold the potential of node A.

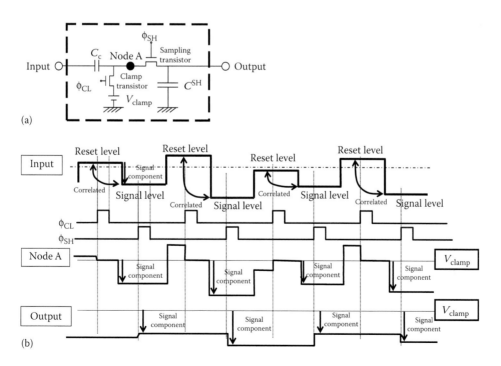

FIGURE 2.27
CDS circuit: (a) configuration; (b) principle operation.

The operation is explained using Figure 2.27b. As for the input, reset level* just after the reset operation of FD and signal level just after the signal charge transfer to FD appear serially in alternate shifts. The signal component that is required is the difference between the reset level and the following signal level of each input signal.

To that end, clamp pulse ϕ_{CL} is applied to the clamp transistor in the reset level period to set the clamp transistor to on-state. This connects node A with clamp voltage V_{clamp}. Clamp operation is completed when the clamp transistor is switched to off-state and node A is set to fixed clamp voltage V_{clamp}. Thus, each potential corresponding to reset level is forced to set clamp voltage V_{clamp} independent of the original reset level in CDS operation. After that, signal charges are transferred to FD, and signal level is output from SFA to CDS input. Only the amount of change (signal component in the figure) of input signal passes through the AC-coupling capacitor C_C to node A to change its potential from V_{clamp} to [V_{clamp} − signal component].

Then, sampling pulse ϕ_{SH} is applied to the gate of the sampling transistor to set it on-state. The potential [V_{clamp} − signal component] is held at the sampling capacitor [C_{SH} and kept till the next signal sampling by setting the sampling transistor off-state. Thus, the CDS circuit can obtain a true signal component by removing the correlated noise involved in both levels with the clamp operation of the reset level and the sampling operation of the signal level. When the sampling operation is achieved at the sampling capacitor C_{SH}, enough electric power is supplied by the SFA.

References

1. W. F. Kosonocky, J. E. Carnes, Basic concept of charge-coupled devices, *RCA Review*, 38, 566–593, 1975.
2. N. Koike, I. Takemoto, K. Sato, A. Sasano, S. Nagahara, M. Kubo, Characteristics of an *npn* structure MOS imager for color camera, *Journal of the Institute of Television Engineers of Japan*, 33(7), 548–553, 1978.
3. W. F. Kosonocky, J. G. Carne, Charge-coupled digital circuit, in *Proceedings of the IEEE International Solid-State Circuits Conference, Digest of Technical Papers*, pp. 162–163, February, 1971.
4. M. White, D. Lampe, F. Blaha, I. Mack, Characterization of surface channel CCD image arrays at low light levels, *IEEE Journal of Solid-State Circuits*, 9(1), 1–12, 1974.

---

* Since each reset level is set at a different voltage because of reset/kTC noise, described in Section 3.2, the variation is emphasized in the figure.

3

Major Types of Noise in Image Sensors

The most important aspect of the performance of image sensors is sensitivity as character-ized by signal-to-noise ratio (SNR). From this perspective, it is important to understand the different types of noise in image sensors. There are various types of image sensors. The main noise type and its level depend on the sensor type, each of which has its own advantages and disadvantages.

Noise disturbs the accurate reception of a signal value by overlapping with it. Various types of noise occur in the time domain (one dimensional), space domain (two dimensional), or both. Noise that fluctuates in the time domain or the time and space domain is called temporal noise, while noise that arises at the same position in images is called fixed-pattern noise (FPN). Since FPN is easy to detect visually, it should be suppressed with high accuracy. On further inspection, there are many examples of noise generated in specific devices or cir-cuitries. Since signals of general image sensors are output sequentially, temporal noise that arises at a common circuit node overlaps with the signal of each pixel and is distributed to the whole image. Although temporal noise appears to fluctuate in the time domain in mov-ing pictures, it becomes FPN in still pictures because there is no time domain.

Noise classification is shown in Figure 3.1. Among the random noise that is a critical issue for image sensors are noise caused by circuitry and transistor devices as well as optical shot noise, which fluctuates in both time and space domains. In recent years, along with shrinkage of the metal-oxide semiconductor (MOS) transistor size, random telegraph noise (RTN), which is generated by particular transistors, has become a serious issue. The level of RTN varies with time at particular pixels and has the property of both temporal noise in the time domain and FPN in the space domain. Therefore, in Figure 3.1, RTN is listed independently from other transistor noise, while it should be categorized as a transistor noise. Since image sensors are basically driven by periodical clock pulses, their leakage causes synchronous noise contamination, which appears at the same position in images and looks like FPN. FPN is also caused by a characteristic variation between the pixels of the image sensors themselves.

Among the types of noise, optical shot noise increases with incident light intensity, as will be described in Section 3.4, and rules the SNR at high light illumination. Other types of noise rule the SNR at low light levels, because they are constant and independent of light intensity.

3.1 Amplitude of Noise

Noise is fluctuation superposed with a signal. Noise varies around a true signal value, as shown in Figure 3.2. Expressing a time-varying signal by noise and a true signal value as $s(t)$ and s_0, respectively, their relation can be shown by

$$\langle s(t) \rangle = s_0 \tag{3.1}$$

	Time Domain (1-D)	Space Domain (2-D)
Random noise	(Temporal noise) Circuit noise Transistor noise RTN	
	Optical shot noise	
Fixed-pattern noise	Synchronous noise	Pixel characteristic variation RTN

FIGURE 3.1
Classification of noise in image sensors.

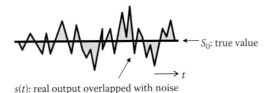

S_0: true value

$s(t)$: real output overlapped with noise

FIGURE 3.2
Signal overlapped with noise and the true signal value along a time axis.

As the amplitude of noise is represented by its difference from a true value, the noise amplitude N is shown as

$$N = \left\langle \left| s(t) - s_0 \right| \right\rangle$$

$$= \sqrt{\left\langle \left| s(t) - s_0 \right|^2 \right\rangle} \tag{3.2}$$

In the case of a complete random temporal noise, the average value $\langle N \rangle$ is zero as follows:

$$\langle N \rangle = \langle s(t) - s_0 \rangle = 0 \tag{3.3}$$

The summation of a number of uncorrelated noises is expressed as follows:

$$N_{\text{total}} = \left\langle \left| \sum_i N_i \right| \right\rangle$$

$$= \left\langle \sqrt{\left| \sum_i N_i \right|^2} \right\rangle \tag{3.4}$$

Using Equation 3.3, Equation 3.4 is rewritten as follows:

$$N_{\text{total}} = \sqrt{\sum_i N_i^2} \tag{3.5}$$

Noise	CCD	MOS	CMOS
Temporal noise	Amplifier noise **Reset noise** **1/f noise** Thermal noise	PD reset noise **kTC noise** **(vertical** **signal line)**	Pixel reset noise Pixel amplification transistor noise **1/f noise** **Thermal noise** **RTN**
	VCCD dark current	**Preamplification transistor noise**	Output amplifer noise
	PD dark current shot noise **Optical shot noise (at high illumination)**		
Fixed-pattern noise	PD dark current (buried PD is effective)		
	Scene noise (nonuniform sensitivity between pixels)		
	VCCD dark current nonuniformity	Column select switching noise	Threshold voltage variability of pixel amplification transistor (effective cancel circuit)

Note: PD, photodiode.

FIGURE 3.3
Noise classification of each type of sensor.

CCD, MOS, and complementary MOS (CMOS) image sensors will be discussed in Chapter 5. The major types of noise are classified in Figure 3.3.

The main temporal noise of a CCD sensor is amplifier noise. An example is reset noise at floating diffusion (FD), which belongs to circuitry noise. Other types of noise are 1/f and thermal noise, both of which are generated at MOS field-effect transistors (MOSFETs). The main types of temporal noise of an MOS sensor are kTC noise of signal lines and preamplifier noise caused at an off-chip transistor. The main types of temporal noise of CMOS sensors are pixel reset noise, 1/f noise of the pixel amplification transistor, and thermal noise and RTN, both of which are transistor noise. Dark current shot noise in a photodiode and optical shot noise are common for the three types.

As for FPN, its source is dark current nonuniformity in CCD. CMOS sensors equip an amplifier in each pixel, and the threshold voltage variation of the amplification transistor causes serious FPN. Although this problem had prevented commercial viability, the barrier was removed by on-chip noise cancelation circuitry, as will be discussed in Section 5.3.

3.2 Circuitry Noise (kTC Noise)

kTC noise is a phenomenon caused at the time of setting a capacitor to a certain voltage by switching it on and off to connect to a voltage source through devices having finite electrical resistance.

In particular, as shown in Figure 3.4a, a capacitor C is connected to and separated from a voltage source V_d through a reset transistor. When the reset pulse in Figure 3.4b is applied to the gate of the reset transistor, ideally the electric potential of node A of one electrode of capacitor C is expected, as shown in Figure 3.4c. That is, node A is settled at voltage V_d excepting only the transition period of the clock pulse. But, in reality, the potential of node A fluctuates around V_d while the reset transistor is on-state, as shown in Figure 3.4d, and node A is settled at the voltage when the reset transistor changes to off-state. Consequently, the potential of node A varies at each reset operation. This is kTC noise and the type of kTC noise caused by the reset operation is called reset noise.

The mechanism is shown schematically in Figure 3.5. The equivalent circuitry is shown in Figure 3.5a, where resistor R means the on-resistance of the reset transistor. Figure 3.5b shows a schematic diagram of the electric charge (electron) distribution while the reset transistor is on-state. The electric charges move about randomly based on Brownian motion by the thermal energy of kT, where T is the absolute temperature. When the reset transistor is on-state, the potential distribution is not uniform because of the nonuniformity of the electric charge distribution by the random motion of the electric charges, since the channel resistance is finite. Here, the readers are expected to imagine a ruffling surface, like a rippling sea, caused by the Brownian motion of electrons. Thus, the fluctuation of a nonuniform potential profile is retained while the reset transistor is on-state. When the transistor changes to off, as shown in Figure 3.5c, the charge quantity existing at node A is retained. Therefore, the potential of node A is set to the potential at that moment.

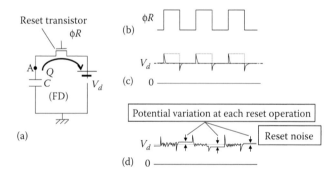

FIGURE 3.4
kTC noise: (a) reset circuitry; (b) reset pulse; (c) potential at node A in an ideal case; (d) potential at node A in an actual case.

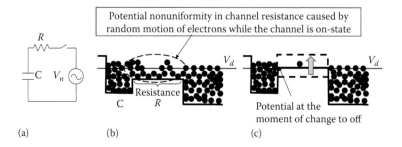

FIGURE 3.5
Generating mechanism of kTC noise: (a) equivalent circuitry; (b) electric charge distribution while reset transistor is on; (c) electric charge distribution at the moment of change to off.

After the transistor turns to off-state, the potential of node A does not change, as node A is floating electrically. Thus, the potential of node A fluctuates at every reset operation. The noise charge quantity, q_n, is expressed as follows:[1]

$$\overline{q_n^2} = kTC\left[1 - \exp(-2t/CR)\right]$$

$$= kTC\,(t \gg CR) \tag{3.6}$$

As shown, this expression is the origin of the name kTC noise.

From Equation 3.6, the noise charge quantity, q_n, and the noise voltage, v_n, are shown as follows:

$$q_n = \sqrt{kTC} \tag{3.7}$$

$$v_n = q_n/C = \sqrt{kT/C} \tag{3.8}$$

Thus, the noise amplitude as an electron number and voltage are directly and inversely proportional to $\sqrt{C}$, respectively.

In the field of image sensors, the amplitude of noise is often expressed by an electron number. Hence, the noise electron number, n_n, at room temperature is expressed as follows:

$$n_n = 400\sqrt{C}\,(C:pF) \tag{3.9}$$

Since the amplitude of kTC noise can be easily estimated, this formula is useful.

Although the noise source of kTC noise exists in the channel of a transistor, the amplitude of noise is determined by circuit construction. Therefore, it is classified as circuit noise.

3.3 Transistor Noise

3.3.1 1/f Noise

Figure 3.6 shows a measured example of the noise power spectrum of a MOSFET with a frequency of 10 MHz. Since the noise power appears inversely proportional to the frequency at less than approximately 100 kHz by plotting on logarithmic coordinates, the noise in this area is called 1/f noise. To date, the mechanism of this noise is unclear, although some models indicate that the charge trap and release at interface states in a Si–SiO$_2$ interface are related. There is an empirical rule that the larger the gate area size that MOSFETs achieve, the lower the 1/f noise power is.

As 1/f noise power is higher in the lower frequency range, the higher-level noise in the lower frequency region overlaps with the image signal, depending on the position of the noise-generating transistor. Since human eyes can follow noise fluctuating at low frequency, this noise is apt to be noticeable as transversal noise visually.

3.3.2 Thermal Noise

The noise spectrum in the higher frequency shown in Figure 3.6 is almost flat and independent from the frequency. The noise in this area is called thermal noise. Noise with a flat

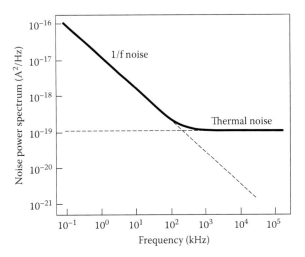

FIGURE 3.6
Measured example of a MOSFET noise power spectrum.

spectrum in a frequency space is also called white noise, meaning that it contains equal amounts of all frequency component. The mechanism of thermal noise originates from the thermal random motion of an electric charge as well as that of kTC noise. Although electric charges flow from the source to the drain through an FET channel as a whole, each electric charge has components of random motion moving in various directions by Brownian motion. Therefore, the electric charge distribution in the channel is inhomogeneous. This means that the density distribution of electric charges that arrive at the drain as current is not homogeneous and varies with time. Accordingly, this inhomogeneous arrival becomes current fluctuations, which are detected as voltage fluctuation via the transconductance of FET, that is, noise. From this mechanism, it is clear that thermal noise is universal for currents flowing through areas of material having nonzero resistance. Thus, although the indications of kTC noise and thermal noise are different, the origin of both is physically the same, that is, the random motion of electrons by thermal energy.

3.3.3 Random Telegraph Noise

RTN is noise in which the channel potential of FETs fluctuates between the quantized states. Figure 3.7 shows a schematic diagram of current fluctuation caused by RTN. Although in many cases of a measured example, the quantized state number is two, there are examples where it is three or four to six.[2] Due to shrinkage of the FET area size, RTN is becoming a serious issue not only for image sensors but also for flash memories. Concerning its mechanism, it is thought that RTN is caused by trapping and releasing an electric charge at the interface state in an Si–SiO$_2$ interface. The indication of RTN is the channel potential fluctuation itself caused by trapping and releasing an electric charge. And it is thought that the two levels correspond to trapped and released states, respectively. The number of interface states that trap an electric charge decreases along with the shrinkage of the gate area size. And it is suspected that the quantized level number is two in the case of the trapping state number being one, and three or four quantized levels correspond to two trapping states. On the other hand, there is a report that the interface states, which are different from those causing 1/f noise, bring about RTN.[3]

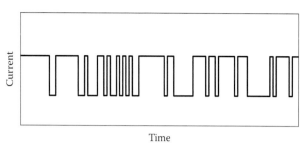

FIGURE 3.7
Pattern diagram of current fluctuation by RTN between two levels.

The image sensors by which RTN is observed are CMOS sensors. These sensors have a transistor in each pixel for amplification, as will be explained in Section 5.3, and among them are particular transistors that cause RTN. Because the outputs of specified FETs fluctuate between multiple levels, they appear in images as a white blemish at a particular position on the pixels.

As a result of recent investigation based on statistical data from numerous MOSFETs with a wide range of sampling times and systems that can measure very low noise, it was clarified that RTN exists even at very low levels (amplitudes), and it was suggested that there is RTN not only in specified groups of FETs, but in every FET. That is, RTN was always observed where that level was not hidden by the noise of the measuring system. In addition, RTN was observed from two states to more than four states. This is quite important information, which dispels the myth that RTN generation is categorized digitally by FET size.

The low-noise system made it possible to observe RTN. This fact evokes the idea from image sensor history that when the origin of the worst noise or white blemish is removed, the second-worst noise or defect (which is behind the worst one) comes out of hiding and so on, iteratively.

3.4 Shot Noise

The physical basis of shot noise originates in the random arrival rate of a discrete particle such as a photon or an electron. The nonuniformity of a photon density distribution in both the time and space domains causes optical shot noise, because photons obey the Poisson distribution. The probability distribution of the Poisson distribution $f(x)$ is shown as follows:

$$f(x) = e^{-\lambda}\lambda^x/x! \tag{3.10}$$

In this distribution, the expected value is the same as with dispersion and is shown as λ in Equation 3.10. The expected value is the average value of the photon number for a signal and is expressed by S. The square root of dispersion corresponds to noise N. The relationship of both is shown as follows:

$$N = \sqrt{S} \tag{3.11}$$

$$S/N = \sqrt{S} \qquad (3.12)$$

It should be noted that the signal charge quantity increases in proportion to the optical intensity, but noise also increases in proportion to the square root of the signal. Thus, the SNR increase is not in direct proportion, but is in proportion, to the square root of the intensity in brighter scenes where the leading noise is optical shot noise.

Dark current shot noise, which is caused by the random generation of thermally excited electrons, also obeys the Poisson distribution. The relationship between the expected value and dispersion is the same as that with optical shot noise.

An example of the illuminance dependence of a signal and noise electron number of a pixel is shown in Figure 3.8. The signal increases in proportion to the illuminance. Dark noise is the summation of noise independent of the illuminance, such as read noise. As shown, shot noise increases in proportion to the square root of illuminance. Overall, noise is the summation of all noise, as shown in Equation 3.5. The SNR is expressed as the logarithmic expression of the ratio of signal to noise in units of decibels as follows:

$$\text{SNR} = 20\log\left(S/N\right) \qquad (3.13)$$

The dynamic range is defined as the logarithmic expression of the ratio of the signal to dark noise in units of decibels as well as the SNR. But this definition is inadequate for nonlinear sensor systems, because the signal is not proportional to illuminance. In such cases, a different expression such as a logarithmic expression of the ratio of the maximum illuminance at saturation to the minimum illuminance at $\text{SNR}=1$ in units of decibels as well as the SNR is required.

Because the noise electron number of recent image sensors is a few at most, the leading noise is optical shot noise, except at low illuminance, which assumes control of the SNR in most cases. It is necessary to realize a highly dynamic range of sensors to achieve truly high-SNR cameras.

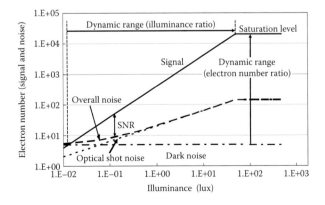

FIGURE 3.8
Illuminance dependence of a signal and noise electron number.

3.5 FDA Noise Reduction by CDS

Among the noise explained, kTC and 1/f noise can be removed or greatly reduced. While the basic operation of an FD amplifier (FDA) was shown in Sections 2.3.1 and 2.3.2, its specific device structure and operation is explained by Figure 3.9.

As shown in Figure 3.9a, the gate electrode input part of a source follower amplifier (SFA) is directly wired with an FD to make up an FDA. It is operated using the following procedure, as shown in Figure 3.9b:

1. Potential of the FD reset to drain voltage V_{RD} is detected and output by the SFA.
2. Signal charge packet is transferred to the FD.
3. Potential of the FD containing the transferred charge packet is detected and output by the SFA.
4. Reset operation to set FD to V_{RD} again by turning the reset gate channel to on-state to accept the next signal charge packet.

At each step after step 4, the potential of the FD is not set to accurate V_{RD}, but varies around V_{RD} because of reset noise, which is kTC noise caused by the reset operation, as explained in Section 3.2. But the potential level of the FD can be detected by the SFA at step 1, that is, the situation with no signal charge in the FD. And by taking the difference between the SFA output at step 3, reset noise can be removed because the SFA outputs at both steps have the exact same reset noise, that is, they are correlated. A representative circuitry to carry out this operation is a correlated double sampling (CDS) circuit, discussed in Section 2.3.3, by making use of the correlation. Thus, a CDS is a circuit to obtain a "true signal without noise" by subtracting (noise) from (true signal + same noise).

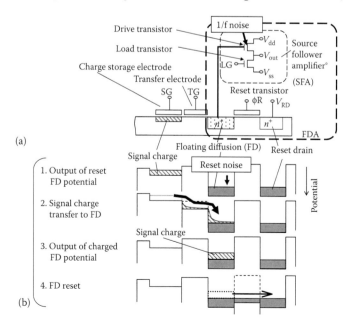

FIGURE 3.9

Operation of FDA and generation of reset and 1/f noise: (a) configuration; (b) operational sequence of FD.

Schematic circuit diagrams of a combination of an FDA and CDS, and that of the operation with an applied clock pulse and potentials and CDS output are shown in Figure 3.10a and b, respectively. Figure 3.10a shows a more practical circuit by introducing a low-pass filter (LPF) to remove higher-frequency components to reduce aliasing, which become a false signal and will be discussed in Section 6.1. An amplifier is also introduced between node A and the sampling part for a higher accuracy of the sampled potential than that shown in Figure 2.26. In Figure 3.10b, ϕR, ϕCL, and ϕSH are the reset pulse, clamp pulse, and sampling pulse, which are applied to the reset transistor, clamp transistor, and sampling transistor, respectively.

The operation starts from the FD reset step, as shown in steps (4) and (1) in Figure 3.9, to set the FD potential to the reset drain voltage V_{RD}. Then, the reset FD potential is detected and output by the SFA in the FDA. After the reset FD potential signal is transmitted to node A, node A is clamped to clamp voltage V_{cl}. Then, the signal charge packet is transferred to the FD as shown in step (2) in Figure 3.9. The potential of the FD changes to the level at which FD contains signal charges, as shown in step (3) in Figure 3.9b, and is detected and output by the SFA as well. Since the FD potential, which is the input of SFA, changes, the output of SFA and necessarily the input to CDS follow. The SFA output amount of change caused by signal charges is transmitted to node A through the AC coupling capacitor C_c. Then, the potential at node A is amplified and sampled at sampling capacitor C_{sh} by the sampling transistor. This is one cycle of CDS operation. Thus, the potential at node A corresponding to the reset level is forced to be clamped to V_{cl} by a clamp pulse after each reset operation, as shown at node A in Figure 3.10b. And the CDS circuit accepts only the change by signal charge packet transfer into the FD. Reset noise causes changes in FD potential by the reset operation, as shown in Figures 3.9b and 3.10a. But any reset noise can be removed by CDS in this way. Outside of reset noise, 1/f noise discussed in Section 3.3.1 occurs at the drive transistor in SFA, as shown by the dashed line at the FDA output in Figure 3.10b. Since its power spectrum is high in the lower-frequency area, the change in the time domain is very slow, as shown in the figure. Therefore, the level of 1/f noise at the time of the clamping and sampling operation can be considered almost equal, because

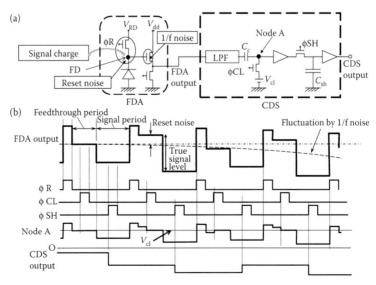

FIGURE 3.10
CDS: (a) circuit configuration of FDA and CDS combination; (b) schematic diagram of CDS operation.

the time difference is very short for 1/f noise to change the level. Then, 1/f noise can be removed by the same operation as CDS as well as reset noise.

Readers might have noticed that both the clamping and sampling operations cause kTC noise. That is correct. But, fortunately, the signal after CDS operation can be treated as a voltage signal, not as a charge quantity signal. As shown in Equation 3.8, the voltage amplitude of kTC noise is inversely proportional to $\sqrt{C}$, thus the noise amplitude problem can be avoided by employing larger capacitors.

References

1. J.E. Carnes, W.F. Kosonocky, Noise source in charge-coupled devices, *RCA Review*, 33, 327–343, 1972.
2. T. Obara, A. Yonezawa, A. Teramoto, R. Kuroda, S. Sugawa, T. Ohmi, Extraction of time constants ratio over nine orders of magnitude for understanding random telegraph noise in metal-oxide-semiconductor field-effect transistors, *Japanese Journal of Applied Physics*, 53, 04EC19, 2014.
3. R. Kuroda, A. Yonezawa, A. Teramoto, T.-L. Li, Y. Tochigi, S. Sugawa, A statistical evaluation of random telegraph noise of in-pixel source follower equivalent surface and buried-channel transistors, *IEEE Transactions on Electron Devices*, 60(10), 3555–3561, 2013.

4

Integration Period and Scanning Mode

In this chapter, scanning modes are described before image sensors are discussed in the next chapter. A description is given of how many separate pieces of information on an image can be used to construct a complete image.

4.1 Progressive Mode

The plainest mode is the progressive mode. As shown in Figure 4.1a, complete image information is constructed by taking image information from individual lines from the first line to the final line by scanning the lines. Each scanning line corresponds to the signal of that line or row of the sensor array. Signals of lines are output in series, as shown in Figure 4.1b. A series of complete image signals is called a frame, and a series of divided subimage signals is called a field. In progressive scanning mode, a frame is formed by one field, therefore the frame information is equal to that of the field. One serial output starts with the first scanning line, that is, the signal of the first line of the pixel array. Before the second scanning line, a horizontal blanking interval (HBI) appears, which was necessary for the electron beam from the end point of the first line to the top of the second line in image pickup tubes and CRT displays. In the present image sensors, mainly vertical driving clocks are applied in HBI to avoid the superposition of synchronous noise caused by their leakage. Hereafter, combinations of scanning line and following HBI appear serially till the final scanning line. A vertical blanking interval (VBI) then follows to complete one frame of scanning. A VBI was necessary for the electron beam to return from the end point of the final line to the top of the first line as well as an HBI. HBIs and VBIs are also necessary for present image sensors as periods to apply driving pulses, because such pulses are overlapped in signal output as induction noise if applied during the signal output period.

4.2 Interlaced Mode

In the interlaced mode, complete image information is divided into subsets. An example of one of the simplest interlaced modes is shown in Figure 4.2.

As shown in Figure 4.2a, complete image information is formed by two images: the first image is made by odd lines and the second image by even lines. Thus, frame information is constructed by the first field and second field, as shown in Figure 4.2b.

Therefore, in the serial output, the first field signal, made by odd line signals, is scanned; after that, the second field signal, made by even line signals, is output. In the case of moving images, this sequence is repeated.

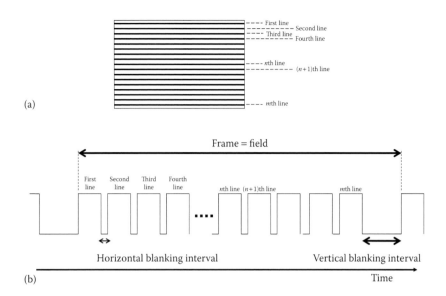

FIGURE 4.1
Progressive mode: (a) progressive scanning lines; (b) serial output and names of each part.

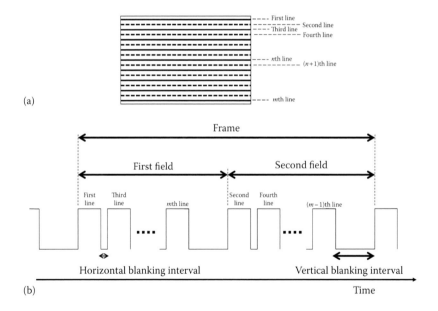

FIGURE 4.2
Interlaced mode: (a) interlaced scanning lines; (b) serial output and names of each part.

Although the above example is interlaced by two fields, there are cases constructed using more fields, depending on the kind of system. Especially, CCDs in digital still cameras have a tendency to employ the interlaced mode with many fields, since many pixels are necessary for high-resolution sensors that have enough time to be operated for readout. The resolution of a still image does not depend on the number of the scanning field, because exposure time is controlled by an optical shutter to capture in progressive mode.

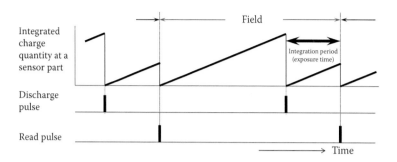

FIGURE 4.3
Operational principle of electronic shutter mode.

4.3 Electronic Shutter Mode

So far, the modes explained have been based on the premise that all signal charges generated by incident light are utilized as signal charges to form image information. But occasionally shorter exposure times are desired by design, because longer time exposures detract from the image sharpness when objects are in motion. Therefore, electronic shutter modes that enable control of exposure times are commonly used. The operational principle is explained by Figure 4.3, which shows integration and discharge of generated signal charges by incident light with time.

As sensor parts are kept illuminated by incident light, integrated photoelectric charges continue to increase with time. But integrated charges can be completely removed by applying a discharge pulse to sensor parts to flush them, as shown. In this way, refreshed sensors restart new integration times until a read pulse is applied. The periods from discharge to readout are real exposure times in which signal charges are integrated and used as image signals. This can be controlled by setting discharge pulse timing. Electronic shutter modes are indispensable for present-day cameras.

5

Types of Image Sensors

Image sensors must have the following functions: (1) a sensor part that converts incident light to signal charges and stores them, (2) a scanning part that identifies the pixel address of each signal charge packet, (3) a part that measures the amount of signal charge and converts it to an electric signal, as discussed in Section 1.3. The choice of devices for these functional elements determines the type of image sensor. From the viewpoint of signal procedure, the order of (2) scanning part and (3) part measuring signal charge quantity may be swapped. Various types of sensors have been proposed in the past.[1] By using technology available at the time, some marketable sensors have been developed. Among them, charged-couple device (CCD), metal-oxide semiconductor (MOS), and complementary metal-oxide semiconductor (CMOS) sensors are representative in terms of greater acceptance in the marketplace. Concrete measures of each functional element of the representative sensor types are shown in Table 5.1.

A pn-junction photodiode (PD) is employed in all types because of its performance and productivity. Thus, a PD is the major sensing element, while a photogate is applied for specific purposes.

In CCD sensors, a CCD is used as the scanning part. Originally, CCD was the name of a device having a charge transfer function, not the name of the image sensor itself. In MOS and CMOS sensors, a shift register or decoder for row/column selection and a metal-oxide semiconductor field-effect transistor (MOSFET) switch are employed as scanning parts to realize X-Y addressing.

A floating diffusion amplifier (FDA) is used as the charge quantity measuring part in almost all CCD and CMOS sensors. The difference is that in CCD sensors the FDA is equipped in a chip as a common amplifier, while in CMOS sensors an FDA is provided at each pixel. Therefore, charge–voltage conversion is carried out at a common FDA in the final stage in CCDs, while in CMOS sensors, this is done at each individual pixel amplifier. In the case of MOS sensors, the voltage signal caused by a minute signal current is detected at an off-chip amplifier.

5.1 CCD Sensors

First the principle of CCDs, then interline transfer CCDs (IT-CCDs), which are the main type of CCD, are described.

5.1.1 Principle of CCDs

The CCD was proposed as a device to store and transfer electric charges by Bell Telephone Laboratories in 1970.[2] The principle is shown in Figure 5.1. One electrode each of adjacently arrayed multiple capacitors are commonly connected to ground level, while other electrodes are independent, as shown in Figure 5.1a. Then, commonly connected

TABLE 5.1

Devices for Each Functional Element in Each Sensor Type

Type	(1) Sensor Part	(2) Scanning Part	(3) Charge Quantity Measuring Part
CCD sensor	*pn*-junction PD, photogate	CCD	FDA /sensor chip (common)
MOS sensor	*pn*-junction PD, photogate	Shift register/decoder and MOSFET switch	JFET amplifier (off-chip amplifier)
CMOS sensor	*pn*-junction PD, photogate	Shift register/decoder and MOSFET switch	FDA/pixel (individual)

electrodes are replaced with one electrode, as in Figure 5.1b. By applying positive voltage to another independent electrode, positive and negative charges are stored in another and one electrode of the capacitor. By shifting one of the independent electrodes to which positive voltage is applied to the next electrode, negative charges stored at the opposite position in the common electrode also shift to the position corresponding to the next electrode, as shown in Figure 5.1b. That is, charges are transferred.

In real devices, the silicon substrate plays the role of the commonly connected electrodes, on which a polycrystalline silicon (polysilicon) layer is formed. This creates further electrodes by sandwiching a gate-insulating film such as silicon dioxide (SiO_2) or silicon nitride (Si_3N_4) between them, as shown in Figure 5.2a.

The electrodes are separated into four groups, in each of which electrodes are connected periodically to all four electrodes, as shown in Figure 5.2a, and to which one of the clock pulses of $\phi1$, $\phi2$, $\phi3$, and $\phi4$ in Figure 5.3 is applied. The operation proceeds as shown in Figure 5.2b. At time t_{1-1}, positive voltage is applied to electrodes $\phi V1$ and $\phi V2$ to store charge packet in the channel potential well of the electrodes, and voltage is also applied to the next electrode, $\phi V3$, to distribute charges to the channel at t_{1-2}. By decreasing the voltage applied to electrode $\phi V1$, the charge packet ranges under $\phi V2$ and $\phi V3$ at time t_{1-3}, that is, charge packet has been transferred by one electrode. Then, the same process is repeated till the charge packet is transferred to a specific position. The mode shown in Figure 5.2 is named four-phase CCD because of the four transfer clock pulses. Since CCD is the capacitor, the maximum charge quantity that can be stored is in proportion to the gate electrode area. In the mode shown in Figure 5.2, the charge packet is stored in at least two electrode channels. This is desirable to increase the maximum charge quantity, which is directly related to dynamic range performance of the sensor. On the other hand, a time length of four clocks from t_{1-1} to t_{1-5} is necessary to transfer charges to the next stage. The time length of eight clocks from t_{1-1} to t_{2-1} is needed to complete the charge transfer of one cycle; therefore this mode is not suitable for high-speed transfer application.

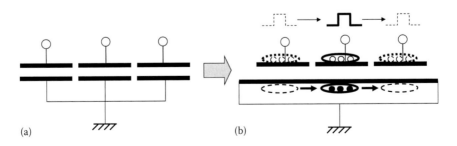

(a)   (b)

FIGURE 5.1

(a,b) Schematic diagram of operational principle of CCD.

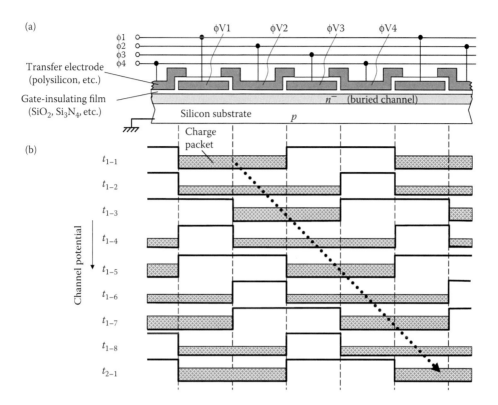

FIGURE 5.2
Four-phase CCD: (a) cross-sectional view; (b) schematic diagram of charge transfer.

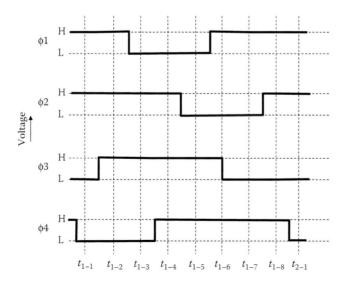

FIGURE 5.3
Driving clock pulse of overlapping four-phase CCD.

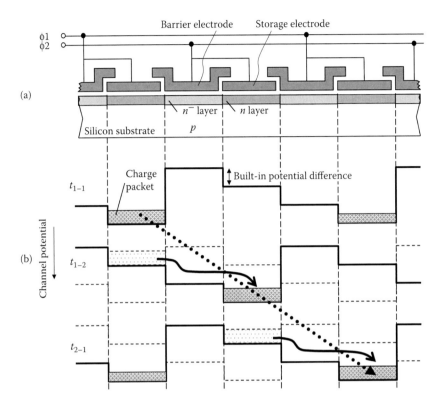

FIGURE 5.4
Two-phase CCD: (a) cross-sectional view; (b) schematic diagram of charge transfer.

The next CCD, which has two phases, is expressed as a proper mode for high-speed transfer. Cross-sectional schematic diagrams of charge transfer and driving clock pulse are shown in Figures 5.4a, b and 5.5, respectively. Two adjacent electrodes are connected and a built-in potential difference is formed between paired electrode channels by introducing different impurity concentrations in each channel. The electrodes with higher and lower channel potentials are called storage and barrier electrodes, respectively. Pair electrodes are driven by application of the same clock pulse, maintaining the built-in potential difference. Using

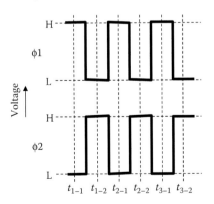

FIGURE 5.5
Driving clock pulse of two-phase CCD.

this structure, transfer direction is defined unambiguously. Charges are stored in the channel of the storage electrode and pass through the barrier electrode channel only in transfer operations. The charge quantity that can be stored is that which fills the built-in potential difference in the storage electrode channel. As shown in Figure 5.5, complementary pulses are applied as driving clock pulses. At time t_{1-1}, the charge packet is stored in a storage electrode channel. Since voltages of $\phi1$ and $\phi2$ switch places at t_{1-2}, charges in the storage gate channel of $\phi1$ are transferred to that of $\phi2$ through the barrier gate channel of $\phi2$. Thus, the charge packet is transferred to the next stage by only one clock pulse time, and one period of time is only two clock pulses. Therefore, the two-phase mode CCD is suitable for high-speed transfer application, although it is not appropriate for use of large charge quantities.

The channel structure used in real CCD sensors is the buried MOS structure mentioned in Section 2.1.4 to achieve high transfer efficiency by avoiding interference of interface states. This is called a buried-channel CCD (BCCD),[3] while CCDs having normal MOS are named surface-channel CCDs (SCCDs). Comparisons of structure and potential distribution between SCCD and BCCD are shown in Figure 5.6a and b.

Since the maximum channel potential (minimum potential energy for electron) of a SCCD is at the surface of silicon, charges are transferred at the interface. Therefore, electrons are trapped at the interface states and released behind them. This phenomenon causes degradation of transfer efficiency. On the other hand, the maximum potential of a BCCD does not reside at the surface but inside the substrate, as explained in Figure 2.14; charges pass through inside the silicon. As charges are transferred without interference by the interface state, a high transfer efficiency can be realized. Currently, CCDs have BCCD channel types. In Figure 5.6c, gate voltage V_g dependence on channel potential ϕ_{ch} for both BCCDs and SCCDs is shown. While the channel potential of SCCD is almost 0 V at gate voltage of 0 V, that of BCCD is above zero because a positive voltage is applied to the n-region to deplete the buried channel. The channel potential varies with gate voltage. But when gate voltage is changed to be larger in the negative direction, holes begin to be collected in the valence band to form a surface inversion layer at the surface at a specific voltage. A further increase of negative voltage is used only to collect more holes, and channel potential is pinned at the voltage. This gate voltage is named pinning voltage. In almost all real CCDs, applied voltage ϕ [H] is 0 V and ϕ [L] is approximately equal to pinning voltage. Therefore, pinning voltage and channel potential ϕ_{ch}^0 at V_g of 0 V are important design parameters.

Why have CCDs cornered the market for so long? The reason is the "complete transfer" of the transfer operation, as transfer charges are highly efficient. Focusing on a specific gate electrode, as shown in Figure 5.7, it receives 100 electrons from the left gate channel and passes on those 100 electrons to the right gate channel, that is, there is neither an increase nor a decrease of electron number. This means no noise is generated in the transfer operation. Therefore, a CCD can achieve a low-noise performance and consequently a high signal-to-noise ratio (SNR), that is, high sensitivity. This is the most important feature of CCDs and the origin of their advantage.

As a device, this means complete depletion and no charge remains after complete charge transfer. This means that there is no channel capacitance after transfer completion, according to the definition of capacitance C as the inverse ratio of potential voltage change ΔV caused by charge quantity change ΔQ. Therefore, kTC noise does not exist. In other words, because no charge is left in the channel, the potential uncertainty formed by a stochastic process does not exist.

A charge transfer device named a bucket-brigade device (BBD), proposed at a similar time, has a similar concept to CCDs, using an MOS-type gate electrode. Charges are

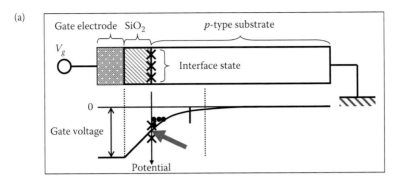

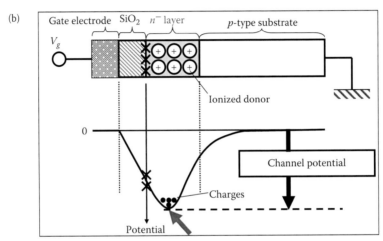

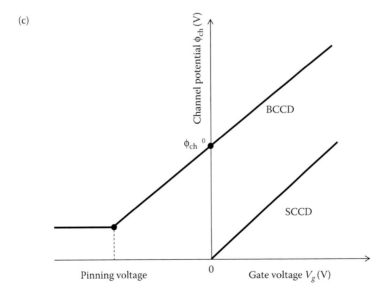

FIGURE 5.6
Comparisons between SCCD and BCCD of (a) structure and potential distribution and (b) gate voltage dependence of channel potential.

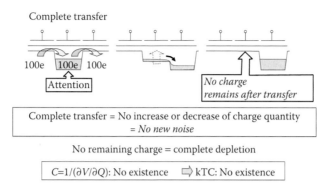

FIGURE 5.7
The significance of complete charge transfer, CCDs' most important feature.

transferred from one MOS channel to another directly in CCDs, while in BBDs they are transferred by way of impurity regions that lie under and between MOS gates and are not depleted. This is related to the technology level at the time they were proposed, the essential difference of the presence of a nondepleted area in BBDs brings about kTC noise and a limit to transfer efficiency.

5.1.1.1 Interline Transfer CCDs

Among image sensors, the only winner for a few decades, since it started to be produced in the mid-1980s in Japan, was the interline transfer CCD (IT-CCD). The first CCD image sensor proposed was the frame transfer CCD (FT-CCD),[4] which has a simpler pixel structure than the IT-CCD. Although the IT-CCD was developed[5] 2 years later than the FT-CCD, it is explained in detail at first regarding the significant role it played.

Figure 5.8 is a schematic diagram of the most common composition of an IT-CCD, having three vertical and three horizontal pixels (3V × 3H).

Optical images are focused on an image area in which pixels are arrayed in a matrix in a plane. The sensor part, consisting of a PD, especially a pinned PD, is equipped in each pixel, as shown in the top right of Figure 5.8. It generates and integrates signal charges by absorbing incident light arriving at the pixel as a portion of the optical image. Each pixel is composed of a sensor part, a transfer electrode to read out signal charges integrated in the sensor part, and a vertical scanner (vertical CCD [VCCD]) to transfer signal charges to vertical direction. Signal charge packets read out to VCCD are transferred through a VCCD to a horizontal scanner (horizontal CCD [HCCD]). Each charge packet arriving at the HCCD is transferred through it to the charge quantity measurement part, consisting of a FDA, and be converted to signal voltage. Thus, scanners of two directions such as vertical and horizontal are necessary in two-dimensional image sensors. A sensor part, transfer electrode, and VCCD are arranged in a narrow area of a pixel. To get higher sensitivity, a larger PD is desired, based on a narrower VCCD area. Therefore, four-phase CCDs, which can treat a larger charge quantity with a narrower area, are preferable for VCCDs, as shown in the middle right of Figure 5.8. Signal charge packets of one line transferred to a HCCD are transferred to a FDA generally at frequencies two or three orders higher than in VCCDs. In contrast to VCCDs, HCCDs have few design limitations regarding width at variance, so it is easy for HCCDs to have

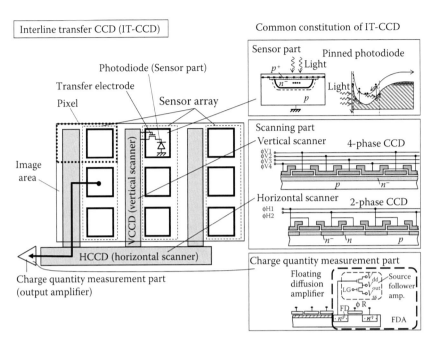

FIGURE 5.8
Schematic diagram of the most common constitution and structure of an IT-CCD.

enough area to deal with the necessary amount of charge packet. Therefore, two-phase CCDs are suitable for HCCDs, which require high-frequency transfer with few design limitations of area. FDAs are used mostly to measure charge quantity. Regions other than sensor parts that receive incident light are often covered with metal layers such as aluminum so as not to be affected by that light. Although *pn*-junction PDs are commonly used as sensor parts nowadays, photogate-type sensors were used in the first IT-CCD proposed.[5]

Figure 5.9 is a schematic diagram of an operation of IT-CCD in progressive mode, explained through time. Figure 5.9a shows a situation where exposure time has finished with stored signal charges in each PD. Signal charges are read out to the VCCD all together, as shown in Figure 5.9b. Signal charges are separated from the sensor part and the next exposure period starts in each sensor part, although this is not shown in the figure. Then charges are transferred in the VCCD toward the HCCD in synchronization and charges of the bottom line are transferred to HCCD, as Figure 5.9c shows. Then, signal charge packets are transferred to the output part to be converted to signal voltages one by one, shown as S11 from pixel address (1, 1) in Figure 5.9d. Each charge packet of the bottom line is transferred to output parts and converted to S11, S12 and S13, which are obtained as in Figure 5.9e. After one line signal is output, all signal charge packets in the VCCD are transferred toward the HCCD and the subsequent line signal charges are transferred to the HCCD, as Figure 5.9f shows. Each charge packet is converted to a signal voltage in the same manner. When the top line signal is output, as shown in Figure 5.9g, one frame output is completed, as shown Figure 5.9h, and signal charge integration for the next frame at each sensor part is completed at the same time. The operation in Figure 5.9a is repeated the necessary number of times.

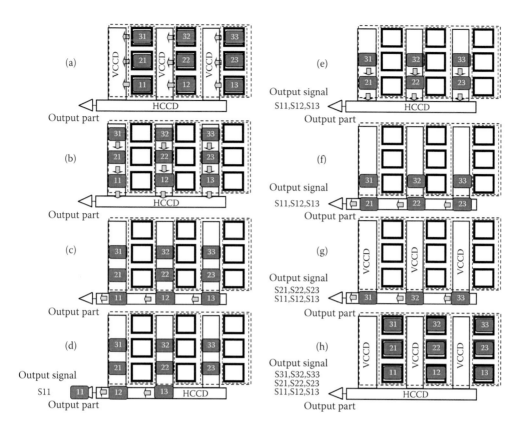

FIGURE 5.9
(a–h) Schematic diagram of IT-CCD operation in progressive mode.

5.1.1.2 Basic Pixel Structure of IT-CCD

A representative pixel structure of IT-CCD is shown as a 3D-like illustration in Figure 5.10. A channel of a BCCD and a pinned PD are formed in the *p*-well built in the *n*-type substrate. The *n*-type region of PD is basically surrounded by a *p*-type region, including two sides of separation between neighboring PDs, although they are not shown explicitly in the figure. The surface side of the PD is covered with a p^+ layer to construct a pinned PD. Since the PD receives incident light, no material that decreases light intensity by absorption or reflection is formed above the PD. On the VCCD channel, two polysilicon transfer gates, ϕV4 and ϕV1, are formed on contact with a gate-insulating layer such as SiO_2. Figure 5.10 shows a representative pixel configuration with two transfer electrodes per pixel. A pixel with two transfer electrodes of ϕV2 and ϕV3 is put on the near side in the transfer direction next to the pixel of the figure, and the two types of pixel are arrayed alternately to form a four-phase CCD. Zero (GND) and positive V-sub biases are applied to *p*-well and *n*-type substrates, respectively. As will be discussed in Section 5.1.2.1, V-sub voltage is set so that excess carriers generated in PDs overflow to the *n*-substrate before they spill over to the VCCD channel or neighboring PDs. Here, gate electrode ϕV1, which is one polysilicon layer, plays the roles of not only a transfer electrode in the VCCD but also a transfer electrode to read out signal charges in PDs to the VCCD channel. Therefore, the electrode is extended to the area above the charge readout path in the figure. Clock pulses applied to electrodes are shown in Figure 5.11 for the interlaced mode. In this case, three-value

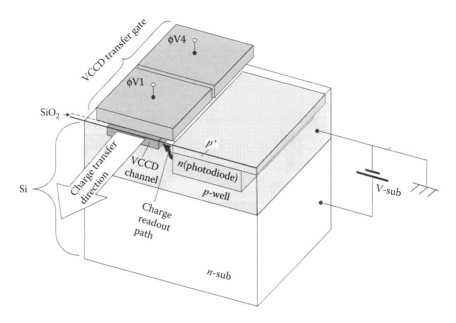

FIGURE 5.10
Pixel structure of an IT-CCD.

voltage clocks having [H, M, L] (about 12, 0, and –5 V, respectively) ϕ1 and ϕ3 are applied to electrodes ϕV1 and ϕV3, and two-value voltage clocks having [H, L] (about 0 and -5V) ϕ2 and ϕ4 are applied to ϕV2 and ϕV4, respectively.

By applying a read pulse of ϕ1 to electrode ϕV1, the VCCD channel potential under electrode ϕV1 in Figure 5.10, which is already depleted as in Figure 5.6b, rises, as shown in Figure 5.6c, and the VCCD channel depletion region is enlarged to the charge readout path.

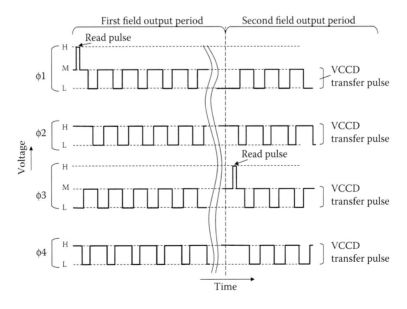

FIGURE 5.11
Readout and transfer clock of a four-phase CCD in interlace mode.

The read pulse is also applied to the portion just above the readout pass of the electrode φV1, which also increases the potential of the charge readout path. Thus, the readout path is efficiently increased by the synergetic effect[6] from both the VCCD channel and surface sides, and signal charges integrated in the PD are read out to the VCCD channel. The readout signal charge packets are transferred in the VCCD channel by applied clock pulses of [M, L] of φ1 and φ3 and [H, L] of φ2 and φ4. Then, signal charge packets corresponding to one-line pixels are transferred to the HCCD and each charge packet is transferred through the HCCD to the output amplifier to be converted to signal voltage output, as shown in Figure 5.9. Once the first field readout is completed, the second field readout starts by applying a read pulse of φ3 to electrode φV3 and is carried out in the same manner as the first field readout. The readouts of these two fields are repeated in alternate shifts. In this mode, odd-line pixel signals are output in the first field readout period, while even lines are output in the second field readout period.

This is a kind of interlaced mode called frame integration mode. It is a typical driving mode for still image capture, with many more field numbers in the current era.

If this mode is applied for moving pictures, a phenomenon named *system lag* occurs, as shown in Figure 5.12a. The sensor has 500 line pixels, that is, 1000 transfer gate electrodes for a four-phase VCCD, and can treat 250 signal charge packets in a one-column VCCD. The first field signal of odd-line pixels and second field signal of even-line pixels, each of 250 lines, are output alternately in 1/60 s. Although the integration period is 1/30 s, images are output every 1/60 s. There is a 50% overlap in the time domain between consecutive fields, as shown in Figure 5.12a; that is, the images with time information 50% the same are successively displayed. This causes an image lag phenomenon known as system lag. Also, there is a driving mode called field integration mode, as shown in Figure 5.12b. In this mode, signal charges of all pixels are read out to the VCCD and signal charges of adjacent

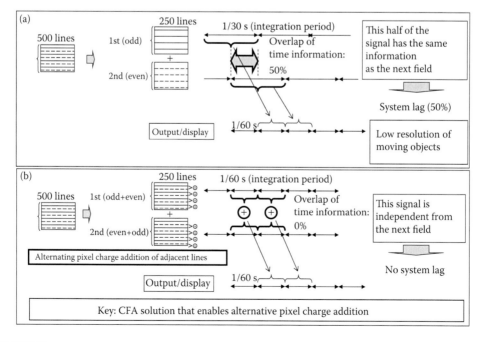

FIGURE 5.12
(a,b) Comparison of integration modes for moving pictures.

odd-line and even-line pixels are added into 250 lines from 500 lines. The combination pair of adding is alternately changed between upper and lower adjacent pixels in each field. Therefore, both the integration period and output period are 1/60 s long without time overlapping between consecutive fields; no system lag occurs.

The above field integration mode is easily applied for monochrome images. But for the case of a single-chip color camera (SCCC) system, which is the widest application, the key is the color filter array solution, which makes it possible to produce a color image by combination with field integration mode. After the key solution of color filter array and system named line sequential color difference system[7] was proposed, the combination of the color filter array and field integration mode became the standard technology of consumer-use SCCCs such as camcorders.

5.1.2 Pixel Technology of IT-CCD

Signal charges are integrated at PDs at first, then read out to the VCCD channel, and transferred through the VCCD and HCCD channel to the output amplifier. As was already mentioned, the most important feature of CCDs is low-noise performance by the grace of complete charge transfer, as shown Figure 5.13a. But if larger level noises are generated in a PD, low-noise performance at the next charge transfer process is not utilized. Therefore, to make effective use of low-noise performance at the transfer stage, a high SNR at the PD stage is indispensable, as shown in Figure 5.13b. As a result, developments to improve SNR performance at pixel level by increasing signal charge quantity S and decreasing noise electrons N are the history of CCD technology.

A cross-sectional structure of that kind of representative pixel technology is shown in Figure 5.14. A silicon substrate of n-type, on which a p-well is formed, with a buried-channel VCCD and pinned PD covered with a p^+ layer, is arranged. The n-region of the PD has low impurity concentration so that it is completely depleted after the signal charges are read out to the VCCD channel. The p-well region just under the PD has a vertical overflow drain (VOD) structure to drain excess carriers to the n-substrate before they spill over into the VCCD channel or peripheral PDs, as mentioned in Section 5.1.1.2. In the upper region

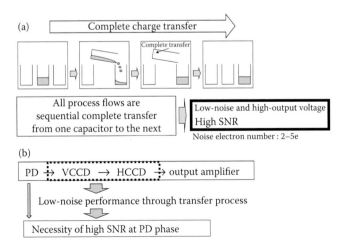

FIGURE 5.13
Features IT-CCDs: (a) low noise by complete charge transfer; (b) necessity of high signal and low-noise-level pixel.

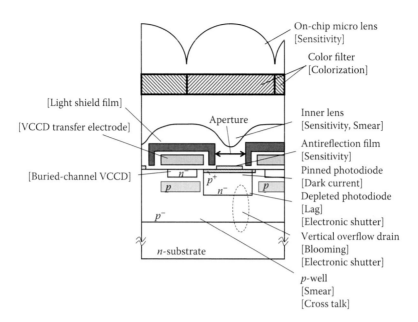

On-chip micro lens [Sensitivity]

Color filter [Colorization]

[Light shield film]

[VCCD transfer electrode]

Aperture

[Buried-channel VCCD]

Inner lens [Sensitivity, Smear]

Antireflection film [Sensitivity]

Pinned photodiode [Dark current]

Depleted photodiode [Lag] [Electronic shutter]

Vertical overflow drain [Blooming] [Electronic shutter]

p-well [Smear] [Cross talk]

n-substrate

FIGURE 5.14
Cross-sectional view of IT-CCD representative pixel technology.

of the silicon substrate, VCCD transfer electrodes are formed on contact with the gate-insulating material above the VCCD channel and covered with metal light-shielding material. Looking from above the PD, an on-chip micro lens (OCL),[8] which effectively focuses incident light toward the aperture area, is formed. If there was no OCL, only incident light coming directly to the aperture area could generate signal charge, although no other light could contribute to sensitivity. A color filter that passes only light belonging to a specific wavelength range to the PD is formed under the OCL. Moreover, an inner lens is arranged to direct incident light in a perpendicular direction to the PD,[9] and an antireflection (AR) film is formed on the PD. A mirror-polished silicon wafer shows a silver-like color and reflects about 30%–40% of visible light. The AR film reduces the reflection at the silicon surface to increase the intensity of the light arriving at the PD. The thickness of the AR film is set so that the phases of two kinds of light, reflecting at the AR surface and silicon surface, have an antiphase relationship to cancel each other's amplitude by the interference effect, using material such as SiO_2 and Si_3N_4.

As mentioned above, an OCL, inner lens, and AR film have been introduced to increase the signal. It is clear that efforts to increase signal S by making much of light showering onto the image sensor lead to PDs with lower loss. The techniques reducing noise N will be shown next.

5.1.2.1 Vertical Overflow Drain Structure

The capacity of each PD is finite, although signal charges are kept generating by continuous incident light and integrated in each PD. If a pixel that is irradiated with high-intensity light generates too much signal charge to be stored in the PD, it overflows to VCCD channels and neighboring PDs. This phenomenon is called *blooming* and is shown in Figure 5.15a. It can be seen that VCCD channels are flooded by excess charges and there is no image information in the area.

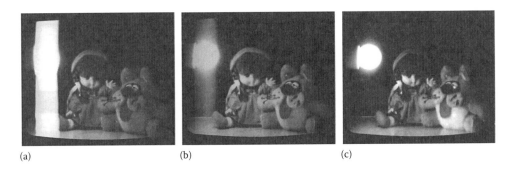

(a)   (b)   (c)

FIGURE 5.15
Images at high-intensity light irradiation: (a) blooming phenomenon; (b) blooming suppression, low smear suppression; (c) blooming and smear suppression.

Figure 5.16 is a schematic diagram of the mechanism of the blooming phenomenon. In practical use, this kind of image collapse has to be avoided. The means for excess charges to discharge is an overflow drain. The first to be proposed is a lateral overflow drain,[10] shown in Figure 5.17. An overflow control gate and overflow drain are formed adjacent to the PD. The channel potential under the overflow control gate is controlled so that excess charges flow to the overflow drain before spilling over to the VCCD channel. Thus, a lateral overflow drain and overflow control gate can suppress blooming and images such as Figure 5.15b are obtained. However, they waste the active area of the silicon surface, because they do not contribute to performance such as sensitivity or dynamic range. They decrease the available area for the PD and VCCD, that is, performance decrement. And that is where the VOD structure is.[11]

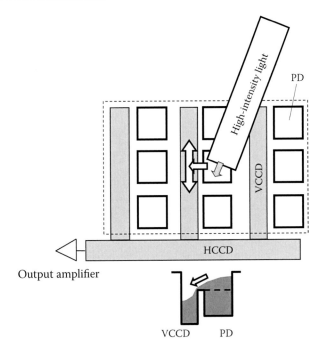

FIGURE 5.16
Mechanism of blooming.

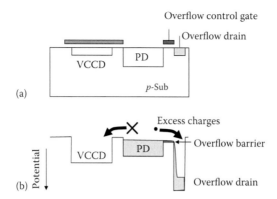

FIGURE 5.17
Lateral overflow drain: (a) cross-sectional view, (b) potential distribution.

As shown in Figure 5.18, the PD and VCCD are arranged in a *p*-well formed on the surface side of the *n*-type substrate. Zero volt or GND potential and positive voltage V-sub are applied to the *p*-well and *n*-substrate, respectively. The potential distribution along the A–A' line in Figure 5.18a is shown in Figure 5.18b. It is important that the *p*-region between the PD and *n*-substrate is completely depleted and the potential can be controlled by applied voltage V-sub, so that excess charges in the PD are discharged to the *n*-substrate before they overflow to the VCCD channel. Since the potential of the overflow control

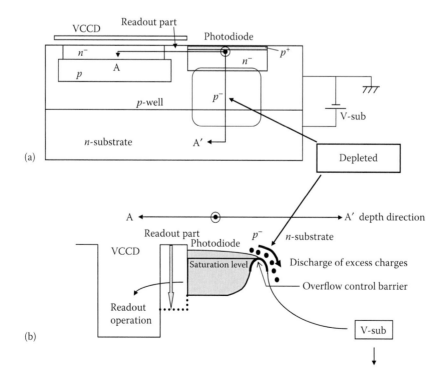

FIGURE 5.18
Vertical overflow drain structure in an IT-CCD: (a) cross-sectional view of the pixel; (b) potential distribution along the A–A' line in (a).

barrier is adjusted by the voltage of V-sub, blooming is not suppressed in the case of a V-sub voltage that is too low, that is, a barrier that is too high, while saturation level is very low in the case of a V-sub voltage that is too high, that is, low dynamic range performance is caused. Therefore, there is an optimal V-sub voltage. But the conditions depend on the impurity concentration of the *n*-substrate, and processing conditions and their variations to form the *p*-well and PD. Therefore, the best voltage is different at each image sensor chip, so adjusted voltage is applied at each chip.

Looking at Figure 5.15b, although the blooming phenomenon is well suppressed, light white vertical bars are seen above and below the lamp. This is a phenomenon named *smear*, caused by incident light or generated charges mixing directly into the VCCD channel, as shown in Figure 5.19.

When image sensors receive high-intensity incident light, many charges are generated around the PDs and smear becomes conspicuous. Since charges generated in the *p*-type substrate diffuse isotropically because there is no electric field, there is a certain probability of them flowing into the VCCD channel. Signal charges in the VCCD channel are transferred to the perpendicular direction of a plane of paper. The quantity of smear charges flowing into VCCD is determined by the product of [structure factor] and [time factor].[12] Structure factor means the ratio of charges flowing into the VCCD channel to total charges generated over a certain length of time and is determined by the pixel structure and electric field distribution in the pixel. On the other hand, time factor means the length of time it takes for the VCCDs to accept the smearing charges determined by the structure factor. Therefore, the smearing charge quantity is inverse proportion to the VCCD transfer frequency.

Generated charges in deep areas of the *p*-type substrate can contribute as signal charges by entering the PD as well as smear charges by flowing into the VCCD channel. As mentioned in Figure 2.21, in the case of *p*-well structure image sensors, charges generated in deeper regions under the dividing ridge of potential are discharged out of the chip through the *n*-type substrate. Since they do not contribute to smearing charges, smear is drastically suppressed. But smear charges entering the VCCD channel through other paths, such as the gap between the shielding film and silicon surface, should be suppressed by different methods, as shown in Figure 5.20.

5.1.2.2 Depleted Photodiode and Transfer Mechanism

In Figure 5.21, the same portion as Figure 5.18b is shown. The situations of signal charges integrated in the PD just before readout operation, early state just after readout operation

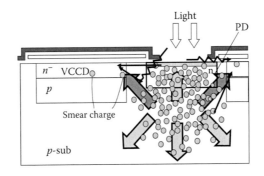

FIGURE 5.19
Mechanism of smear phenomenon at cross-sectional view of pixel. Generated charges are flowing in. Light enters the VCCD channel directly.

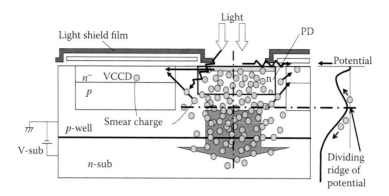

FIGURE 5.20
Smear suppression mechanism in *p*-well structure image sensors. Generated charges in regions deeper than the dividing ridge of potential are discharged through the *n*-type substrate.

start, final state of readout operation with little charges left, and completed state of readout operation are shown in Figure 5.21a–d, respectively. On readout operation, the read pulse shown in Figure 5.11 is applied to the transfer gates, which play roles as not only transfer gates of the VCCD but also as readout transfer gates from the PD to the VCCD channel. Then, a readout channel potential rises from ϕ_{off} to ϕ_{read}, as shown in Figure 5.21b. ϕ_{PD}, ϕ_{off}, and ϕ_{read} are completely depleted PD potential, readout channel potential at OFF, and READOUT situation, respectively. The potential of PD is increased until all the charges are transferred to the VCCD channel and the PD is completely depleted, as shown in Figure 5.21d.

Here we trace the process of charge transfer from start to finish. In Figure 5.21b, the readout channel has just been opened and many charges are still in the PD. Therefore, almost all charges are transferred by self-induced drift, that is, mutual coulomb repulsion between charges. In the final state of transfer operation, as in Figure 5.21c, since charge quantity is very small, coulomb repulsion is not a leading mechanism of transfer anymore. If there is graduation of potential, shown by the broken line in Figure 5.21c, that is, an electric field toward the VCCD channel exists, remaining charges are transferred to the VCCD channel quickly and smoothly. But if there is no electric field, shown by the straight line in Figure 5.21c, each charge moves in a random direction based on thermal motion. Since there is no direction to move, only charges that happen to move to the VCCD channel direction and fall into it are transferred to the VCCD channel. It takes a long time for all the charges to be transferred by falling in by chance and thus completing transfer to the situation in Figure 5.18d. This causes a profound difference in the length of time necessary to complete transfer.

For complete charge transfer, it is clear that the potential of ϕ_{read} should be higher than that of ϕ_{PD}. A situation where there is a potential barrier or a dip in the charge transfer path, as shown in Figure 5.22a, must be avoided. As shown in Figure 5.22b, the PD, readout channel, and VCCD channel can be considered as sources, with voltage V_s, gate of V_g, and drain of V_d of a MOSFET. Let us consider the transfer operation as a behavior of a MOSFET. In the initial situation of transfer, because $V_g - V_s \gg kT$, channel current I_s is expressed as follows:

$$I_s \propto \left(V_g - V_s\right)^2 \tag{5.1}$$

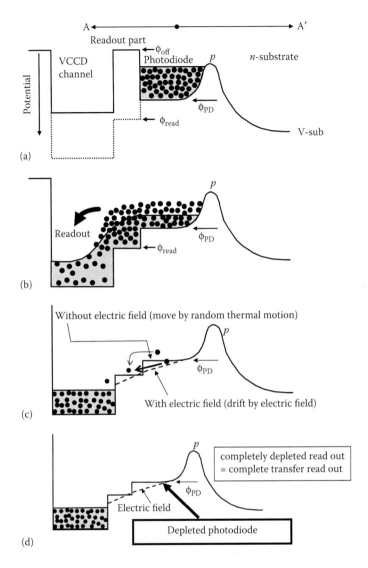

FIGURE 5.21
Schematic diagram of transfer mechanism in depleted PD: (a) before readout; (b) early state of readout operation; (c) final state of readout process; (d) after completion of readout.

and the charge amount in the PD decreases rapidly. But if there is a potential barrier, as shown in Figure 5.22b, at the stage that charge quantity reduces to the level that $(V_g - V_s)$ decreases to the same degree with several-fold of kT, I_s decreases very much as follows:

$$I_s \propto \exp\left[q\left(V_g - V_s\right)\right]/kT - 1 \tag{5.2}$$

This is the weak inversion mode or subthreshold region of a MOSFET.[13] In this situation, the leading mechanism of electron movement is thermoelectron emission mode, and it

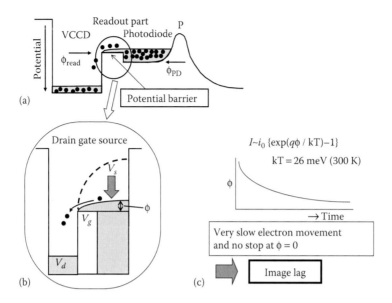

FIGURE 5.22
Schematic view of charge transfer in a nondepleted PD: (a) potential distribution; (b) weak inversion mode in MOSFET operation model; (c) time dependence of current by thermoelectron emission.

takes a very long time to reach equilibrium condition, although V_s does not stop decreasing at $V_s = V_g$. As a behavior of image sensors, all the signal charges cannot be read out in a field and are read out throughout many fields. This is image lag in motion pictures. Since complete charge transfer is achieved in depleted PDs, image lag does not occur. kTC noise does not occur either, as mentioned in Section 3.2.

While the example of incomplete charge transfer described here is a case of the charge transfer from the PD to VCCD channel, the mechanism is the same for charge transfer in the CCD channel.

Another operation mode that needs a completely depleted PD is electronic shutter mode, mentioned in Section 4.3. In electronic shutter operation in VOD structure image sensors, all the signal charges integrated in the PDs are flushed to the substrate at the same time. After a prescribed integration period, signal charges in the PDs are read out to the VCCD channel.

As shown in Figure 5.23a, the discharge operation is carried out by applying a high positive voltage pulse to the *n*-type substrate to push down the *p*-region potential barrier so that charges in the PD are flushed to the substrate. As shown in Figure 5.23b, discharge and readout operations have different paths. If there is excess or insufficient charge quantity at discharge or readout operation, fixed-pattern noise (FPN) is caused. Therefore, it is necessary that the PD situations just after both discharge and readout operations are perfectly equal.[14] Since the potential of the PD is determined to be fixed depleted potential after discharge or readout operation in the case of a fully depleted PD, this kind of problem does not occur.[15] To achieve this effect, it is necessary that both readout and discharge potentials are higher than the depleted potential of the PD.

The discharge and readout operations in MOS image sensors, which will be discussed in Section 5.2, are not done in transfer mode but in reset mode, by a direct connection, so there is no such problem. The three-transistor pixel configuration in the CMOS image sensor, which will be mentioned in Section 5.3, has the same reset operation as that of the

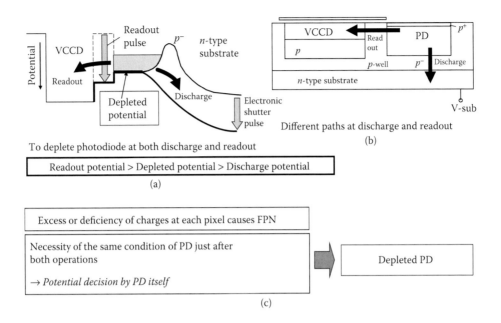

FIGURE 5.23
Schematic diagram of electronic shutter operation at the completely depleted PD in a VOD structure: (a) potential distribution; (b) paths for discharge and readout; (c) necessity of completely depleted PD.

MOS sensor. There is no similar problem. But the four-transistor pixel configuration in the CMOS sensor has the same transfer operation as that of CCD, so complete transfer is required at readout operation.

Looking at Figures 5.21 and 5.22, some readers might consider the transfer and movement of electrons as analogous with water flow. But this is not correct. Water movement is controlled by gravitation and surface tension. On the other hand, electron movement is ruled by electric field and thermal energy. The two mechanisms are quite different. Readers need to break away from the "water model" to understand this accurately.

5.1.2.3 Buried Photodiode/Pinned Photodiode

The structure of a buried PD[13] or pinned PD[16] was shown in Section 2.1.7, although the functions were not discussed in detail. A PD can exist that is buried but not pinned, and one that is pinned but not buried can also exist. Buried PDs are named after their structure, while pinned PDs are so called because their potential is pinned by complete depletion. A buried/pinned PD was proposed to prevent lag phenomena due to incomplete transfer, which was seen in normal *pn*-junction PDs in early IT-CCDs. Therefore, their initial purpose was to achieve a lag-free PD. The function of dark current suppression was not strongly focused on, because dark current caused by *pn*-PDs was not a severe problem for the level of CCDs at the time. So it is possible that the existence of lag phenomena incubated buried/pinned PDs.

Before discussing this, let us consider the normal *pn*-junction PD shown in Figure 5.24. There is a depletion layer at the junction of the *n*- and *p*-type impurity areas. It also exists at the silicon surface, where interface states exist. These interface states mostly play the role of stepping stone of thermally excited electrons from valence band to conduction band, as shown in the energy diagram along the interface in Figure 5.24b; that is, charges

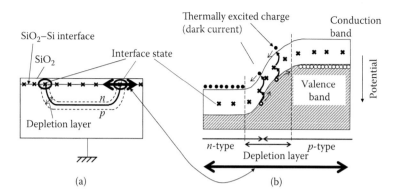

FIGURE 5.24
pn-Junction PD: (a) cross-sectional view of structure; (b) energy diagram along interface including depletion layer.

are generated even under dark conditions, especially in the depletion layer, as mentioned below. Therefore, they are called dark current.

Expressing n and p as electron and hole density, respectively, there is a rule in semiconductors that the product np is a constant value decided by the absolute temperature under equilibrium conditions as a result of the balance between generation and recombination. Since this does not depend on n/p-type, it is the same with the square of intrinsic carrier density n_i. The value of n_i is $1.5 \times 10^{10}/cm^3$ in the case of silicon at room temperature. Therefore, phenomena tend toward the thermal equilibrium in which the product np is $2.3 \times 10^{20}/cm^3$. The impurity concentration of a nondepleted region is at the level of 10^{14}–$10^{17}/cm^3$, and carrier density is at the same degree. But in a depleted region, the densities of electrons in the conduction band and holes in the valence band are very low. The product np is near zero. Because this situation is the farthest one from equilibrium, the speed of change toward equilibrium is very high. Therefore, charge generation rate is very high. In nondepleted regions under nonequilibrium, charges are also excited toward equilibrium by using the interface state as stepping stones. But the rate is not so high because there are charges with the density of impurity concentration mentioned above. On the other hand, the dark current by way of interface state in depleted regions is predominant. Therefore, in the case of a PD fully depleted to the whole surface, many charges are thermally generated, that is, there is a very high level of dark current.

From the above discussion, it can be said that it is effective for dark current suppression to bring about near-equilibrium conditions at the surface to decrease the density of additional electrons by increasing the carrier density around the interface. Theoretically, either electrons or holes are possible. But it is necessary to increase hole density rather than electron density to be consistent with realization of the depleted PD and completely cover the depletion layer at the interface.

A high-concentration p-type layer in the range of 10^{17}–$10^{19}/cm^3$ is introduced at the surface of the PD in buried/pinned PDs, as shown in Figure 5.25. Therefore, almost the same density of holes exists in this space. Hence, the density of electrons that can exist around the interface is reduced to the level of 10–$10^3/cm^3$. Thus, the thermal excitation probability by way of interface state is very low, and generation of dark current is drastically suppressed. It is desirable that the impurity concentration of the p^+ layer is high from the viewpoint of dark current suppression. And the thickness should be thin to avoid sensitivity loss caused by recombination of signal electrons with high density holes. To compare dark

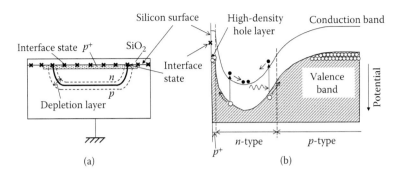

FIGURE 5.25
Buried/pinned PD: (a) cross-sectional view; (b) energy diagram.

currents of a normal photodiode (n^+p-PD) and pinned photodiode (p^+np-PD), photographs taken at 73°C with a 2 s exposure period are shown in Figure 5.26. Image quality degradation caused by FPN originating from variation of dark current is strongly seen in normal PDs. By using buried PDs, dark current can be reduced to less than one-tenth of that of a normal PD.

As discussed above, one role of the p^+ layer is to suppress dark current. But it has another role, to stabilize the potential of the PD and suppress the potential increase of the depleted PD by pulling down from the surface side as well as the bottom side. If a PD without a high surface concentration p-type layer is fully depleted, its potential is affected by the charges in and on the SiO_2 layer on the PD and the electric line of force radiating from neighboring gates, although a pinned PD is realized. Since electric potentials of gates change with time, the potential of a PD is unstable. It causes the increase of readout pulse voltage for complete transfer readout. As the surface p^+ layer is connected with GND level by way of the p-well, this kind of instability is avoided.

Buried/pinned PDs are widely used for high-performance sensors also in CMOS image sensors because of low dark current, low-noise, and image lag–free performance.

From the operational principle of pinned PDs by complete depletion, it is clear that the maximum signal electron number of a PD is the same as the n-type impurity atom number at the highest. However, in practice it is a lower number because of the limitations placed

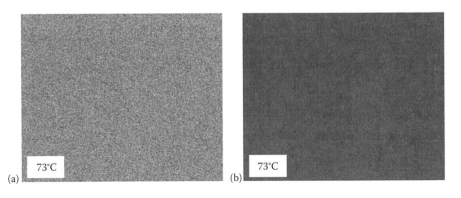

FIGURE 5.26
Comparison of dark current images at 2 s exposure at 73°C: (a) normal PD (n^+p-PD); (b) buried/pinned PD (p^+np-PD).

on the available potential range by blooming suppression. This limitation is fast approaching as pixel size becomes progressively smaller.

5.1.3 Progress of CCD Sensor

In this section, variation and progress of CCDs are described.

5.1.3.1 Frame Transfer CCD

FT-CCDs were proposed as the first CCD-type image sensor[4] rather than IT-CCDs. As shown in Figure 5.27, an FT-CCD with 3(V) × 3(H) pixels is composed of four blocks: image area, storage area (which does not exist in IT-CCDs), horizontal shift register (HCCD), and output amplifier. In the imaging area, pixels are arrayed in a matrix in a plane. All parts except for image area are usually covered with a light shield film so that they are not affected by incident light. A four-phase CCD and isolation are built into the image area and storage area. In the case of IT-CCDs, the sensor part and vertical shift register (VCCD) are arranged independently. In contrast, there is no structural distinction in FT-CCDs; the two functions work by time-sharing. During the exposure period, the CCD in the image area integrates signal charges as sensor part. In the frameshift period, which is just after the exposure period, it transfers signal charges integrated in it to the CCD in the storage area with high frequency as vertical shift register (VCCD). Cross-sectional views and potential distributions of pixels along vertical and horizontal directions are shown in the

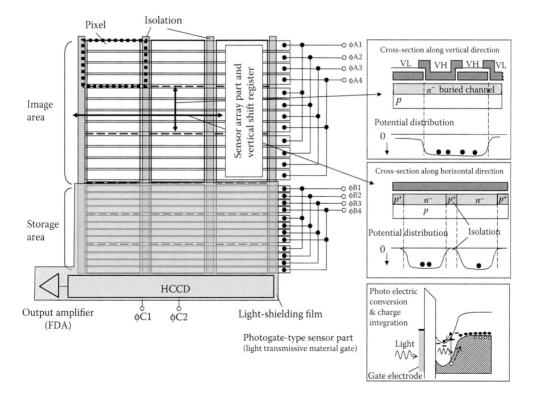

FIGURE 5.27
Configuration and structure of FT-CCD.

upper and central right parts of Figure 5.27. In this example, low voltage V_L is applied to two neighboring gates and higher voltage V_H is applied to the other two adjacent gate electrodes to form a potential well to integrate signal charges during the exposure period. While it works as a sensor part, incident light is irradiated on the gate electrodes, as shown in the right bottom part of Figure 5.27; that is, it is a photogate-type sensor. Although poly-crystal silicon is mainly used as transfer electrodes in CCDs, it is also a kind of silicon. Therefore, the spectral response is similar to that of crystal silicon. As can be understood from penetration depth in Figure 2.20, light transmittance is high enough if the gate electrode is thinner than 0.1 μm. But in real situations, it is difficult to realize thin enough electrodes because they should have sufficiently low electrical resistance to be driven in high frequency. So the sensitivity is apt to be lower under the 500 nm wavelength region.

The operation is explained by Figure 5.28. Figure 5.28a shows the situation where the exposure period has just finished and signal charges have been integrated in all pixels labeled 11–33. All the signal charges are transferred to the storage area at high speed simultaneously, as shown in Figure 5.28b. Because whole charges are transferred as one frame, this transfer is called frameshift, and is the origin of the name of frame transfer CCD. By this transfer, charges are separated from the sensor part (imaging area), and signal integration for the next exposure starts, although this is not shown in the figure. This situation corresponds to Figure 5.9b in IT-CCD. Hereafter, the sequence where signal charges in the storage area are transferred to the HCCD line by line is shown in Figure 5.28c. Each charge packet in the HCCD is transferred to the output part, which is the FDA, one by one to be converted to voltage signal output, such as S11 shown in Figure 5.28d. This is the same situation as IT-CCD in Figure 5.9d. Thus, each line is transferred to the HCCD and output serially. One frame output is completed at Figure 5.28f, and signal integration for the next frame is also completed. In the case of interlace drive mode, two pairs of adjacent electrodes alternate, forming a potential well.

Some part of light or charges come in the VCCD when there is smear in IT-CCDs, but what about in FT-CCDs? During exposure, as signal charges being transferred for output are in the storage area, which is covered with a light shield film, smear does not occur. But in the frameshift periods, the image area is receiving incident light as well as in the exposure periods, that is, all generated charges become smear charges, so the structure factor is 100%. The only thing that can be done is to reduce the time factor. Therefore, frameshift in high frequency is necessary to shorten the time factor. There have been pro-posals to block out incident light using a mechanical or electro-optical shutters to avoid smear phenomenon completely during the frameshift period.

Although pixels have a simple structure composed of only CCD and isolation, FT-CCDs need an additional storage area, such as frame memory, separate from the image area, because pixels work as sensors and transfer CCDs at each different operating time, and the two operations are not concomitant in pixels. This causes an increase in sensor die size. On the other hand, in the case of IT-CCDs, since the VCCD in the pixel plays the role of frame memory, no additional storage area is necessary, although the pixel structure is more complicated than that of FT-CCDs. From the viewpoint of production, this means that IT-CCDs need finer processing technologies, although FT-CCDs need larger die size by avoiding fine processing.

5.1.3.2 *Frame-Interline Transfer CCD*

In this section, we discuss the frame-interline transfer CCD (FIT-CCD),[17] which was the standard sensor for applications requiring very high image quality such as studio cameras for broadcasting. An FIT-CCD is constituted by installing a storage area like an FT-CCD

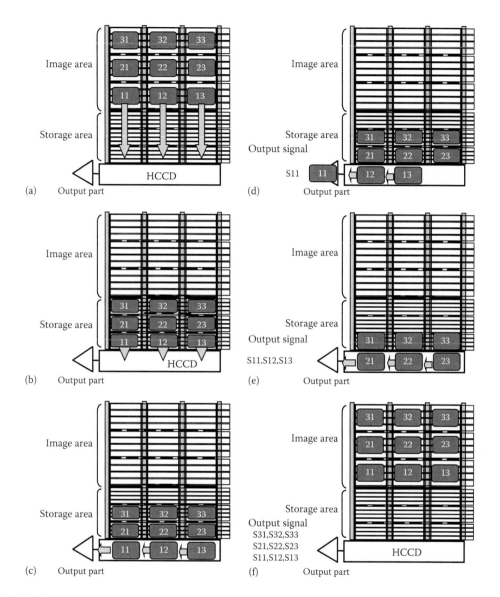

FIGURE 5.28
(a–f) Schematic diagram of FT-CCD operation.

between the image area and horizontal shift register (HCCD) of an IT-CCD, as shown in Figure 5.29.

The operation to read out signal charges integrated in PDs to the VCCD is the same as in IT-CCDs. Frameshift, in which charges in the VCCD are transferred to the storage area at high frequency, follows the same as in FT-CCDs. Charge packets in the storage area are transferred to the HCCD line by line, and transferred charge packets are converted to voltage signal output at the output amplifier one by one, in the same manner as in IT-CCDs and FT-CCDs. The major purpose of this type of sensor is to achieve a very low smear level. By high-speed frameshift, charge packets pass through the image area being irradiated in a short time. This reduces the time factor of smear phenomenon. While the vertical

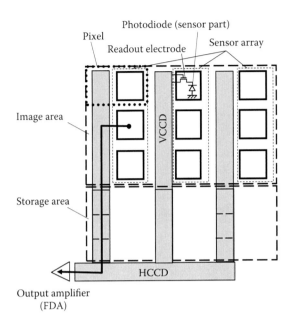

FIGURE 5.29
Device configuration of FIT-CCD.

transfer frequency of IT-CCDs is around 10 KHz, the transfer frequency of FIT-CCDs is around 1 MHz or more, that is, one hundredth or less the smear level of IT-CCDs while obtaining the same pixel structure. Thus, an image sensor with a very low smear level was developed and was employed as a standard sensor type for broadcasting cameras for a long time.

5.2 MOS Sensors

MOS image sensors are XY-address-type sensors.[18,19] They were mass-produced for consumer-use video cameras by Hitachi for the first time as solid state image sensors in 1981.[20]

5.2.1 Principle of MOS Sensors

In the case of CCDs, addressing is done by signal charges transfer through the VCCD and HCCD to the output part. In contrast, in MOS image sensors this is realized by selecting vertical and horizontal MOSFET switches. Consequently, charges flow through a metal signal line to the output part as signal current, as shown in Figure 5.30a with only one pixel.[21]

Vertical selection (row selection) and horizontal selection (column selection) are done by the output pulses of vertical and horizontal shift registers, respectively. When signal charges are read out, first by switching on the vertical MOSFET switch, charges move to the vertical signal line because of the very large capacitor ratio of vertical signal line to PD. Second, horizontal selection pulse is applied to the switch MOSFET to connect the vertical signal line to video voltage through the horizontal signal line. Since signal current is in

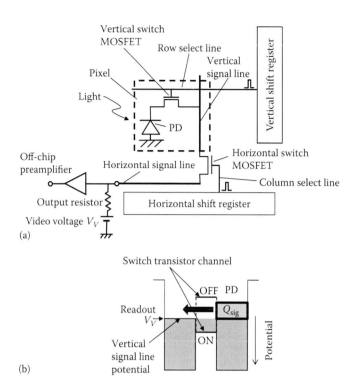

FIGURE 5.30
Operational principle of MOS image sensor: (a) basic constitution; (b) switching mode by MOSFET.

proportion to signal charge quantity, voltage depression caused by signal current across the output resistor is detected as output voltage by the preamplifier. In this example, the signal charge is electrons, so signal charges flow from the PD to the video voltage source at output operation, while current flows in the opposite direction.

Described in terms of electric circuitry, the PD is charged to video voltage V_V by being connected to the video voltage source through vertical and horizontal signal lines and being electrically cut off to be in floating condition to start signal charge integration for the next readout. During the exposure period, the potential of the PD decreases gradually by discharge accompanied by integration of signal charges generated by incident light, and is recharged to V_V again by the next readout operation. The amount of signal current compensates for the quantity of charge discharged by incident light. The voltage drop caused by discharging current across the output resistor is detected by the off-chip pre-amplifier. The maximum current caused by signal charges are generally some hundreds of nanoamperes (nA) at most. And an output resistor with very high resistance cannot be used because of the required frequency performance. Therefore, the amount of output voltage obtained is about one thousandth or less of the output voltage of CCDs obtained by the source follower amplifier (SFA). Thus, the current readout method is disadvantageous compared to voltage readout using a SFA from the viewpoint of SNR.

It should be noted that the readout mode from the PD to vertical signal line is based on switching mode, in which the potential level of the switch transistor channel is higher than those of both the PD and vertical signal line, as shown in Figure 5.30b. It is quite different from the charge transfer mode of IT-CCDs accompanied with complete depletion, as shown in Figure 5.21. Hitherto, the operational principle is described by focusing on only one PD.

Figure 5.31a is a schematic diagram of the image area of an MOS sensor. As shown in the figure, the image area is made up of two dimensionally arrayed pixels, each of which has one PD, a vertical switch MOSFET, and parts of a row select line and vertical signal line. For vertical scanning, the output part of the vertical shift register, which selects a line to read out, called *accessing*, is connected to the row select line to transmit an output pulse to the vertical switch MOSFET to read out signal charges in the PD. Horizontal scanning, horizontal shift register, column select line, horizontal switch MOSFET, and horizontal signal line are arranged the same as in vertical operation. The horizontal signal line is connected to the video voltage source by way of output resistor and off-chip pre-amplifier. Operation of the scanning circuit (shift register or decoder), which selects a row or a column to access, is shown in Figure 5.31b. This is an example of a shift register. The shift register is activated by a start pulse, which is not shown in the figure at first. Then the first pulse of the driving clock pulse ϕ_{cl} is applied to the input part, and a pulse to select a row or column appears at Output 1, although no pulse is generated at other outputs. The next clock input pulse causes a pulse at Output 2. Thus, the output part at which a pulse appears shifts one by one at every pulse of driving clock pulse ϕ_{cl}. As the output position shifts in series in the shift-register circuit, all outputs need to be scanned. On the other hand, any output can be accessed in the decoder circuitry. Although the circuit is larger and more input clock numbers are necessary, this type of circuitry is employed for applications that need flexible access function such as partial readout.

Operation of MOS sensors is described in Figure 5.32, starting with Figure 5.32a. First, as shown in Figure 5.32b, a row select pulse generated by the vertical shift register is applied to vertical switches in the first row by way of the first row select line. Since the vertical switches turn to on-state, signal charges integrated in PDs on the first lines are moved to each corresponding vertical signal line in the mode shown in Figure 5.30b. The potential of the PD just before readout has been reduced from video voltage V_V set at the previous readout by an amount of the quantity of integrated signal charges. Since the ranges of capacitance of the PD C_{PD} and vertical signal line C_{VSL} are generally around some femtofarads (fF)

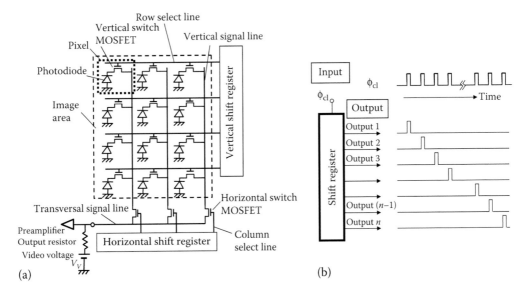

(a)   (b)

FIGURE 5.31
MOS image sensor: (a) basic configuration; (b) operation of shift register.

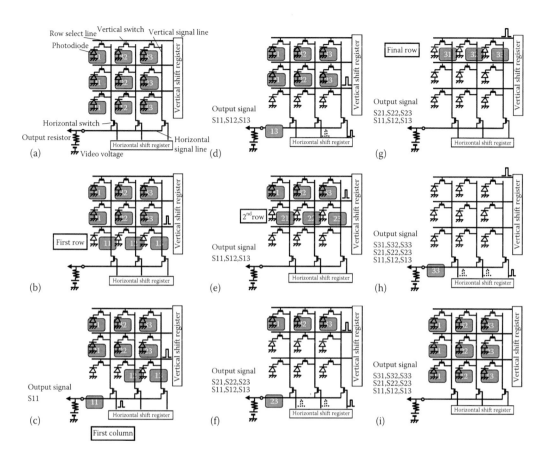

FIGURE 5.32
(a–i) Operation of MOS image sensor.

and ten picofarads (pF), respectively, that is, $C_{PD} \ll C_{VSL}$, almost all the signal charges move to the vertical signal line by redistribution due to the on-state of the vertical switch. Next, a column select pulse generated by the horizontal shift register is applied to the horizontal switch of the first column, as shown in Figure 5.32c. Then, signal charge packet [11] from the PD at the first row and first column flows to video voltage source by way of output resistor. The voltage caused by voltage drop of signal current across the resistor is detected by the off-chip preamplifier as voltage output signal S11. Subsequently, readout of the first row is completed by signal charges [12] and [13] are output as output signals S12 and S13, as shown in Figure 5.32d. Figure 5.32e and f show the readout from pixels in the second row. By the output of the final row signal, a whole readout sequence of one field is completed, as shown in Figure 5.32g–i.

Turning back to the readout operation of the MOS sensor, when the vertical switch is turned off, the potential of the PD is set to video voltage V_V. And when the horizontal switch is turned off after an output signal from one pixel is obtained, the potential of the vertical signal line is also set to video voltage V_V. These operations are the same as the reset operation that causes kTC noise, as discussed in Section 3.2.1.

Supposing capacitances of the PD and vertical signal line are 3 fF and 10 pF, respectively, as shown in Figure 5.33, a schematic diagram of the signal charge readout path in the MOS sensor, electron numbers of kTC noise of the PD and vertical signal line can be estimated to

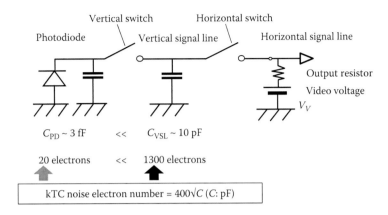

FIGURE 5.33
Capacitance and quantities of kTC noise through the charge readout path in an MOS sensor.

be 20 and 1300 electrons, respectively, by Equation 3.9. The latter is a very large noise electron number. Since the horizontal signal line is kept connected to the video voltage source, it causes no kTC noise. Compared to the fact that no noise is generated by complete charge transfer in CCDs, the kTC noise of the vertical signal line is a major disadvantage of MOS sensors.

As we can see, there is a big difference in exposure timing between CCD and MOS sensors. In IT-CCDs, all PDs are accessed at the same time. All pixels start and complete the exposure period in the same time period in CCDs, as shown in Figure 5.34a. On the other hand, in MOS sensors, each row is accessed one by one in series, that is, the exposure period timing of each line shifts serially, as shown in Figure 5.34b. The same is true of CMOS sensors, as will be discussed in Section 5.3.

5.2.2 Pixel Technology of MOS Sensors

As a characteristic pixel technology of MOS sensors, the *npn*-PD,[22] which is an antiblooming device, is shown in Figure 5.35. The PD is arranged in the *p*-well formed on the *n*-type substrate. Although it seems to resemble the VOD structure in IT-CCDs, discussed in Section 5.1.2.1, the way it suppresses blooming phenomenon is different. In this structure, the *p*-well is not

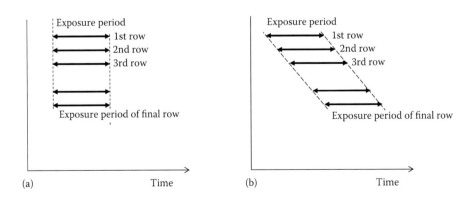

FIGURE 5.34
Comparison of readout timing and exposure period: (a) CCD, all the same exposure timing; (b) MOS/CMOS sensor, exposure timing shift of each row.

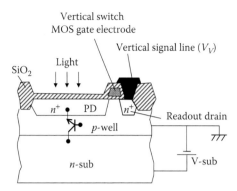

FIGURE 5.35

npn-PD in MOS sensor pixel structure. (Reprinted with permission from Aoki, M., Ando, H., Ohba, S., Takemoto, I., Nagahara, S., Nakano, T., Kubo, M., and Fujita, T., *Transactions on Electron Devices*, ED-29, 745–750, 1982.)

depleted. A bipolar transistor is made up of the PD, *p*-well, and *n*-sub, which play the roles of emitter, base, and collector, respectively. By readout operation, the PD is set to video voltage V_V, which is positive voltage, and starts to integrate signal charges, reducing the PD potential. When excess carriers are stored in the PD, its potential drops to negative voltage, corresponding to the built-in potential (0.6 V) passing through 0 V. The *pn*-junction formed by the emitter and base (PD and *p*-well) becomes a forward-biased condition, and additional excess charges in the PD are emitted to the collector (*n*-sub) by way of the base (*p*-well). This means that current flows from the collector (*n*-sub) to the emitter via the base. Blooming phenomenon is suppressed by this mechanism in MOS sensors, while in the VOD structure of IT-CCDs, excess charges are discharged to the *n*-substrate by depleting the *p*-well, as shown in Figure 5.18. This works similarly to a static induction transistor (SIT). In the *npn*-PD, it works as a bipolar transistor without depletion of the *p*-well. Therefore, the process of adjusting substrate voltage at each sensor is not necessary in production, unlike in IT-CCDs.

It also should be pointed out that the *npn*-PD has the function to reduce sensitivity for far-infrared light, which is excrescence for color image capturing, as discussed in Section 2.2.2.

Smear in CCD arises from the inflow of some part of generated charges or incident light to the VCCD channel while potential wells move through from top to bottom in it. Smear phenomenon also exists in MOS sensors, but the situation is a little different to that in CCDs. In MOS sensors, one vertical signal line is connected to the readout drains of a number of vertical pixels. If some part of generated charges flow into the readout drains, they are not distinguished from signal charges from PDs and are output together, that is, they become smear charges. As the vertical signal line is reset to video voltage at every readout operation of each line, smear charge integration period is one-line readout time. But all readout drains connected to one vertical signal line contribute to smear phenomenon.

5.2.3 Progress in MOS Sensors

There has been some progress in MOS sensors, as follows.

5.2.3.1 Pixel Interpolation Array Imager

In the second generation of MOS sensors, the pixel interpolation array imager shown in Figure 5.36 was developed.[23] The difference is that pixels are arranged with a half pixel

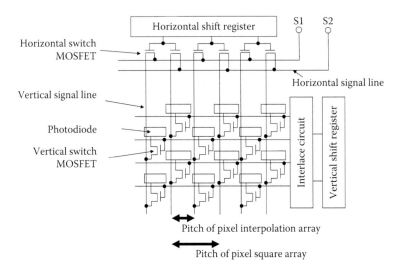

FIGURE 5.36
Device configuration of pixel interpolation array MOS sensor.

pitch displacement in the horizontal direction every two lines, while pixels are in square array in the first-generation sensor, similar to that shown in Figure 5.31a. In the sensors, two adjacent rows are read out simultaneously, the same as in the first generation. Therefore, there are twice as many sampling points in the horizontal direction as in the square array sensor. This means enhancement of horizontal resolution. This is effective not only in monochrome images, but also in SCCCs because of the employed color filter array[24] in which a green component is contained at every pixel site.

5.2.3.2 Transversal Signal Line Imager

The third generation of MOS sensors are transversal signal line (TSL) imagers.[21,25] As shown in Figure 5.33, the major noise in MOS sensors is kTC noise of the vertical signal line. As the device configuration in Figure 5.37 shows, the vertical signal line is not formed in the TSL. Signal charges are read out from the PD to the horizontal signal line directly through vertical and horizontal MOSFET switches formed in each pixel. In the figure, V_R, R_P, S_V, C_V, C_H, and R_f mean reset voltage, reset pulse, signal row select MOS switch, vertical line capacitance, horizontal signal line capacitance, and output resistor, respectively. Using this method, the source of large kTC noise is removed. The remaining kTC noise of PD is about 20 electrons. Since readout drains that collect smear charges are connected to the horizontal signal line, the smear charge integration period is one pixel readout time, that is, smear is reduced to one thousandth. Although TSL imagers need some complications, such as both vertical and horizontal switch MOSFETs in each pixel and both addressing wiring to each pixel, it has achieved drastic performance advances by extreme reduction of sensor noise itself and smear, which is virtually absent. But larger random noise, of around 300 electrons, which is caused in the off-chip preamplifier originating from the current readout mode, remains. As a result, MOS sensors disappeared in the early 1990s, because they could not compete against CCDs, which entered the market in the mid-1980s with lower noise of about 10 electrons at that time. But MOS sensors did not completely vanish; they came back in the late 1990s as CMOS sensors.

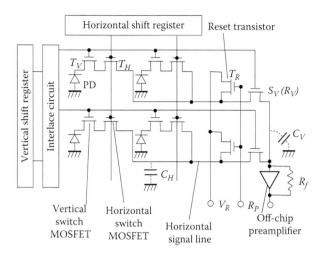

FIGURE 5.37
Device configuration of TSL-type MOS sensor.

5.3 CMOS Sensors

The mass production of CMOS sensors for full-scale digital still cameras was started in 1997 by Toshiba. Although the CMOS sensor can be considered an "improved MOS sensor," a certain level of fine size transistor technology, 0.35 µm, was necessary for commercial viability because of the larger number of elements in a pixel, as will be discussed.

An ancestral CMOS image sensor was proposed by P. Noble in 1968.[26] This was investigated by F. Ando at NHK from around 1980,[27] and his work is internationally recognized as the first substantive CMOS sensor.[28,29]

5.3.1 Principle of CMOS Sensors

The basic concept of CMOS sensors is that although high-level noise is added at the vertical signal line in MOS sensors, the impact of added noise can be suppressed by previous signal amplification. To this end, a pixel amplifier is arranged between the PD and vertical signal line, as shown in Figure 5.38 compared to Figure 5.30a. The amplified signal is treated by MOS switches and metal wirings as well as MOS sensors.

SFAs are most commonly used as pixel amplifiers. While a drive transistor (amplifying transistor) is formed in each pixel, only one common load transistor is arranged for each vertical signal line or column to constitute an SFA. As mentioned in Section 2.3.2, while the voltage gain of an SFA is less than unity, charge quantity is multiplied by the ratio of output capacitance to input capacitance of the SFA. The actual ratio in CMOS sensors is around 100–10,000. By this multiplication, the impacts of noise in later-stage operation, such as kTC noise, which is a fatal disadvantage of MOS sensors, are drastically suppressed by a reciprocal factor of substantive charge quantity gain. This is the most important feature enabling CMOS sensors to compete against CCDs in sensitivity as SNR. The most important feature for pixel-level amplification sensors is the multiplication of charge amount with little additional noise.

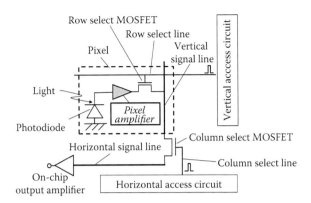

FIGURE 5.38
Basic configuration of CMOS image sensor.

It is also important that CMOS sensors are manufactured based on the CMOS process.*
This means various functional circuitries constructed of CMOS transistors can be arranged
on a sensor chip. Additionally, CMOS sensors can share in the bounty of progress in CMOS
process technology. These features are the origin of evolvability of CMOS sensors.

From the viewpoint of the functional elements discussed in Figure 1.11, the path that signal charges follow at each pixel is in the order of (1) sensor part, (3) signal charge quantity
measuring part, and (2) scanning part, while in CCD and MOS sensors it is in the order of
(1), (2), and (3).

Image sensors with the capability to amplify function in each pixel are called *active pixel
sensors* (APS). Hence, image sensors without this pixel amplifying function, such as CCD
and MOS sensors, are called *passive pixel sensors* (PPS).

5.3.2 Pixel Technology of CMOS Sensors

Since CMOS sensors have an amplifier in each pixel, some elements are necessary in a
pixel. Examples of specific pixel configuration are shown below.

5.3.2.1 Three-Transistor Pixel Configuration

A basic pixel configuration is shown in Figure 5.39. Because of the three transistors in a
pixel, it is called a three-transistor (3-Tr) pixel configuration. The pixel corresponds to the
pixel in Figure 5.38. The *n*-type region of the PD is directly connected to the gate input of
the drive transistor of the SFA, and to voltage source V_{dd} by way of reset transistor RST.
This node has effective input capacitance C_{in}. The source of the drive transistor, which is
the output of the SFA, is connected to the TSL by way of row select transistor RS. The load
transistor is arranged as a common one to form the SFA with drive transistors belonging to
the vertical signal line at one end of the signal line. A vertical signal line is connected to a
number of vertical pixels that have output capacitance C_{sl} of some or some tens of picofarads (pF). Row select transistor RS and reset transistor RST work as switching transistors,
while the drive transistor works as an analog transistor in the SFA. This configuration can

---

* This does not mean that CMOS sensors manufactured by the low-cost vanilla CMOS process can compete
  against CCDs on performance, as will be discussed later.

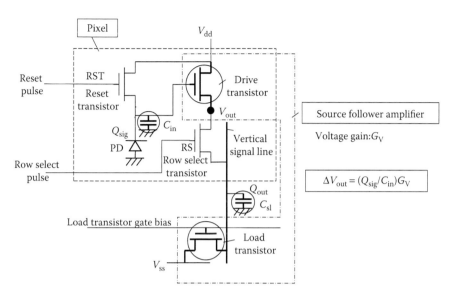

FIGURE 5.39
Three-transistor pixel configuration.

be seen as the FDA forming with drive and load transistors discussed in Sections 2.3.1 and 2.3.2 by regarding C_{in} as C_{dif}. Generated charges Q_{sig} are integrated in the n-region of the PD, whose potential is initially reset to $V_{in}^0 (= V_{dd})$, and resulting potential V_{in}^{sig} is expressed by the following equation, where Q_{sig} includes polarity of signal charge, that is, Q_{sig} should be negative in the case of electrons:

$$V_{in}^{sig} = V_{in}^0 + \frac{Q_{sig}}{C_{in}} \tag{5.3}*$$

Thus, the potential of the n-region changes from V_{in}^0 to V_{in}^{sig} by generated signal charges. Expressing voltage gain of the SFA as $_GV$; the gate threshold voltage of the drive transistor as $_{Vt}h$; amount of voltage change of n-region (input amplitude) as $\Delta_{Vi}n$; and SFA outputs as V_{out}^0 and V_{out}^{sig} at inputs V_{in}^0 and V_{in}^{sig}, respectively, the amount of voltage change of output (output amplitude) and stored signal charge quantity Q_{out} at output capacitance C_{sl} are shown as follows:

$$V_{out}^0 = \left(V_{in}^0 - V_{th} \right) G_V \tag{5.4}$$

$$V_{out}^{sig} = \left(V_{in}^{sig} - V_{th} \right) G_V = \left(V_{in}^0 + \frac{Q_{sig}}{C_{in}} - V_{th} \right) G_V \tag{5.5}$$

$$\Delta V_{in} = V_{in}^{sig} - V_{in}^0 = \frac{Q_{sig}}{C_{in}} \tag{5.6}$$

---

* Since this expression includes polarity of charge, it is $[+Q_{sig}]$. If the signal charge is electrons, then $[Q_{sig}]$ should be negative.

$$\Delta V_{\text{out}} = V_{\text{out}}^{\text{sig}} - V_{\text{out}}^0 = \frac{Q_{\text{sig}}}{C_{\text{in}}} G_V \tag{5.7}$$

$$Q_{\text{out}} = \Delta V_{\text{out}} C_{\text{sl}} = \frac{Q_{\text{sig}}}{C_{\text{in}}} G_V C_{\text{sl}} \tag{5.8}$$

Then, charge quantity gain G_Q is expressed as follows:

$$G_Q = \frac{Q_{\text{out}}}{Q_{\text{sig}}} = \frac{C_{\text{sl}}}{C_{\text{in}}} G_V \tag{5.9}$$

Thus, the charge quantity gain is a product of capacitance ratio of output to input and voltage gain. As general parameters, C_{in} is around some femtofarads, C_{out} is some to some tens of picofarads, and G_V is about 0.7–0.9 V. Hence, quantity of signal charges generated by incident light is multiplied 100–10,000 times. This is the origin of CMOS sensor performance.

When a row to which the pixel to be read out belongs is accessed, the row select pulse generated by the vertical access circuit (vertical shift register or decoder) is applied to the row select transistor by way of the row select line to make it on-state to activate the SFA. After V_{out} is output from the SFA, the reset pulse generated by the vertical access circuit is applied to reset transistor RST to reset the PD to V_{in}^0, by discharging signal charges stored in the n-type region of the PD to source voltage V_{dd}. And signal charge integration for the next exposure starts.

As shown by Equation 5.5, the SFA output voltage depends on the V_{th} of the drive transistor. Threshold voltage has a range of variation of 10–100 mV in a whole chip implemented by general large-scale integration (LSI) process technology. Since the maximum signal level is around some hundreds of millivolts, the direct usage of SFA output V_{out} makes images with dreadful FPN of SNR less than 30 dB, reflecting V_{th} variation. This level is never available for general applications. This phenomenon is one of the reasons for this technology's delayed practical usage.

By replacing V_{in}^0 with V_{dd} in Equation 5.5, we obtain the following equation, where Q_{sig} is substituted by $-Q_{\text{sig}}$ to enhance a practical case of electron signal:

$$V_{\text{out}}^{\text{sig}} = -\frac{Q_{\text{sig}}}{C_{\text{in}}} G_V + \left(V_{\text{dd}} - V_{\text{th}}\right) G_V \tag{5.10}$$

A graph of $V_{\text{out}}^{\text{sig}}$ with variable Q_{sig} is shown in Figure 5.40. Output voltage starting at zero signal charge is $(V_{\text{dd}} - V_{\text{th}})G_V$, and it decreases linearly with increasing electron Q_{sig}, with a gradient of $-G_V/C_{\text{in}}$. The gradient means sensitivity and varies in direct and inverse proportion to G_V and C_{in}, respectively. Since the vertical axis of this line varies depending on the value of V_{th}, an example is shown by the broken line in Figure 5.41. Because this is caused by offset variation, it can be canceled by taking the difference between output voltages with and without signal charges at each pixel. That is, FPN caused by V_{th} variance can be eliminated. Thus, each pixel has its own line according to each different V_{th} value.

Major examples of offset variation canceling circuits are shown in Figure 5.42. Figure 5.42a shows a differential circuit.[30] Output voltages with and without signal charges are stored by sampling at capacitances C_s and C_0, respectively, and difference

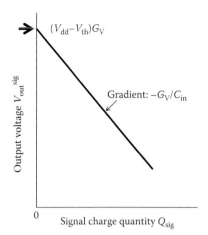

FIGURE 5.40
Signal charge quantity dependence of output voltage of a pixel in the case of an electron.

voltages are obtained by a differential circuit. Figure 5.42b shows a correlated double sampling (CDS) circuit,[31] discussed in Section 2.3.3.

In recent small pixel image sensors, it is hard to arrange a differential circuit as an independent circuit in each column because of the larger circuit size and the narrow column pitch. Therefore, a differential circuit is used as a common circuit at the sensor output or off-chip differential circuit, although high-frequency driving is necessary. In this case, two signal outputs with and without signal charges, having a band frequency around 10–30 MHz, have to be independently transmitted, so twice the original pixel signal transmitting rate is necessary.

On the other hand, CDS circuits are so much smaller that it is possible to arrange one at each column; therefore, canceled signals can be obtained at each column. A canceling operation is carried out at each column by parallel processing within a narrow bandwidth <1 MHz. This is a big advantage from the viewpoint of noise characteristic. So CDS circuits are employed more than differential circuits for offsetting cancelation. The operational method used for offset cancelation in the CMOS sensor,[32] which was mass-produced for

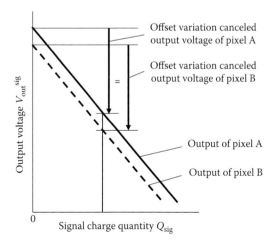

FIGURE 5.41
Signal charge quantity dependence of output voltage of two pixels in the case of electron.

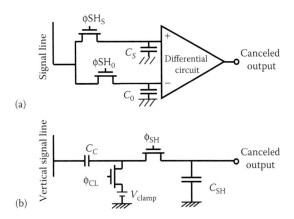

FIGURE 5.42
Examples of an offset variation canceling circuit: (a) differential circuit; (b) CDS circuit.

full-fledged digital still cameras (DSCs) for the first time by Toshiba, is offset cancelation in the charge domain by a cancelation circuit formed at each column,[33] which is similar to the CDS circuit method.

Variations caused by the arrangement of an amplifier at each pixel are not only threshold voltage, as shown in Equation 5.10, but gradient, which voltage sensitivity also varies. Contrary to expectation, it is very rare for variation of voltage sensitivity to be a problem. Because variation of the threshold voltage is randomly determined by microscopic conditions at each transistor channel, correlation between adjacent pixels is low and differences are large, so image quality is degraded drastically. On the other hand, it is thought that mutual conductance g_m, which is seen in the equation of G_V, and the summed capacitances of pn-junction, gate electrode and stray capacitance C_{in}, which mainly rules the voltage gain of the SFA, are macroscopic parameters determined as averaged factors. Since they have a certain level of correlation between adjacent pixels, they are not very different from each other. Moreover, the voltage gain of the SFA also has a character that tends to compensate for the influence of g_m variation.

A schematic diagram of a 3-Tr pixel CMOS sensor is shown in Figure 5.43. In operation, the row selection pulse generated by the vertical access circuit is applied to row select transistor RS as well as the MOS sensor. This makes the SFA active, and the vertical signal line is set to the voltage of the SFA output. Multiplied charge quantity corresponding to the voltage is generated on it. Thus, amplified signal charges in the vertical signal line are supplied by the SFA. This is quite different from the MOS sensor, in which signal charges on the vertical signal line are optically generated signal charges themselves in the PD. Since signal charges in the PD do not move but remain at readout operation, this type of readout is called nondestructive readout. To start the next exposure, the reset pulse generated by the vertical access circuit is applied to reset transistor RS to reset the PD. Then the column selection pulse generated by the horizontal access circuit is applied to column select transistor CS. Potential of the vertical signal line is output by the output part, after the offset variation is canceled by the FPN cancelation circuit. Columns are accessed to be output in series. After all columns in one row are output, the next row is accessed as well as the MOS sensor.

Figure 5.44 is a schematic diagram of a readout operation of one pixel in a 3-Tr pixel CMOS sensor. Figure 5.44a and b show cross-sectional view and potential distribution in

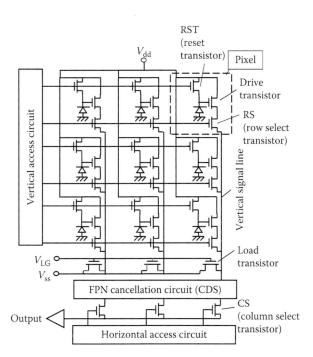

FIGURE 5.43
Schematic diagram of a 3-Tr pixel configuration CMOS sensor.

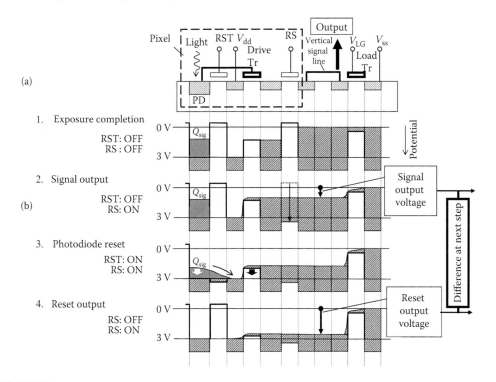

FIGURE 5.44
Operational diagram of one pixel in a 3-Tr configuration: (a) cross-sectional view; (b) operational diagram by potential distribution.

operation, respectively. The source voltages V_{dd} and V_{ss} are set at 3 V and 0 V, respectively. At (1), the exposure period has just been completed, and signal charge packet Q_{sig} is stored in the PD. Next, the row select pulse is applied to the row select transistor RS to activate the SFA, in which electrons start to flow from V_{ss} to V_{dd}, as shown in (2). Here the SFA outputs signal output voltage, because the PD holds Q_{sig} in it and is connected to the gate electrode of the drive transistor. After the output, the reset pulse is applied to reset transistor RST to reset the PD to voltage V_{dd} at (3). Consequently, the SFA outputs the voltage signal corresponding to the PD potential reset to V_{dd}. Offset cancelation is achieved at the next step by taking the difference of outputs (2) and (4).

As discussed above, 3-Tr pixel is a basic configuration of CMOS sensors. But currently, this configuration is not employed for CMOS sensors, which can compete with CCDs on performance. The reason is that the 3-Tr pixel configuration has the following three disadvantages: (1) Dark current is higher. Because the *n*-type region of the PD is directly connected with metal wiring as a line to a gate of the drive transistor, a buried PD cannot be applied. (2) It is difficult to realize higher output voltage. Since the PD also plays the role of floating diffusion (FD), its capacitance cannot be made low enough to achieve a sufficiently dynamic range. In other words, it is difficult to make the gradient in Figure 5.40 high. (3) The reset (kTC) noise of the PD cannot be canceled, because of different reset situations of signal output and reset output, that is, there is no correlation between two reset noises, since a reset operation exists between them, as shown Figure 5.44. A frame memory is necessary for storage of previous reset conditions to realize reset noise cancelation. Of course, offset variation is removed by cancel circuits.

To overcome these disadvantages, 4-Tr pixel configuration is employed for high-performance CMOS sensors, as will be discussed in Section 5.3.2.2.

5.3.2.2 Four-Transistor Pixel Configuration

Three-transistor and four-transistor pixels are compared in Figure 5.45. The difference in the 4-Tr pixel configuration is that a diffusion capacitance FD, which is independent of the PD, and readout transistor TX, which controls the readout transfer of signal charges from PD to FD, are introduced between the PD and drive transistor of the 3-Tr pixel

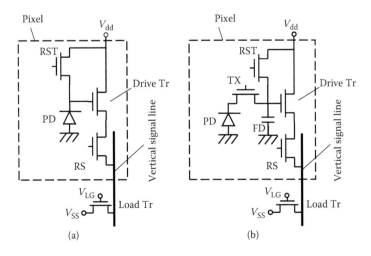

FIGURE 5.45
Comparison of pixel configuration: (a) 3-Tr pixel; (b) 4-Tr pixel.

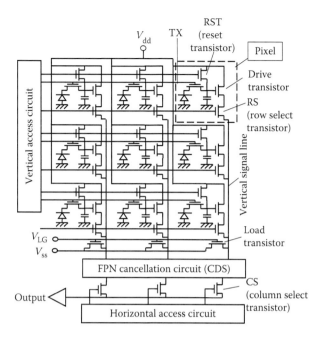

FIGURE 5.46
Schematic diagram of a 4-Tr pixel configuration CMOS sensor.

configuration. In readout operation, signal charges in the PD are transferred to FD by applying readout pulse to TX, and potential change of FD is detected by the SFA, which is made up of a drive transistor and load transistor as well as the 3-Tr pixel.

Figure 5.46 is a schematic diagram of a 4-Tr pixel configuration CMOS sensor. Overall, the operation is the same as that of the 3-Tr sensor except for signal charge readout transfer from PD to FD and reset operations caused by division of roles between PD and FD.

Figure 5.47 is a schematic diagram of the readout operation of one pixel in a 4-Tr pixel CMOS sensor. Figure 5.47a and b show cross-sectional view and potential distribution in operation, respectively. In Figure 5.47b, (1) shows just after exposure period completion and signal charges Q_{sig} and noise charges Q_{noise} are integrated in the PD and FD, respectively. At first, the SFA is activated by applying the row select pulse to row select transistor RS and electrons start to flow from V_{ss} to V_{dd}, as shown in (2). Before readout operation, noise charges Q_{noise} integrated in FD are reset also, as shown in (2). Then, the reset level of FD is output as reset output voltage, as shown in (3). Signal charges Q_{sig} integrated in the PD are transferred to FD by applying readout pulse generated by the vertical access circuit to the gate of readout transfer transistor TX, as shown in (4). It should be noted that this is the charge transfer operation mentioned in Section 5.1.2.2. The potential of FD is decrease by the signal charge packet and the level is output as signal output voltage by SFA, as shown in (5). After signal level is output, signal charges in FD are reset, as shown in (6). Then one readout and output sequence is completed, as shown in (7). The next exposure period starts when signal charge transfer is completed, as shown in (5).

The difference of signal output voltage and reset output voltage is obtained at the next stage by an offset cancelation circuit. Because the reset situations of the FD of both of them are the same, not only offset variation but also kTC noise of FD can be removed. Because the time length between reset operation of FD and signal charge output is quite short, the amount of integrated dark current at FD is negligible. Since no direct contact

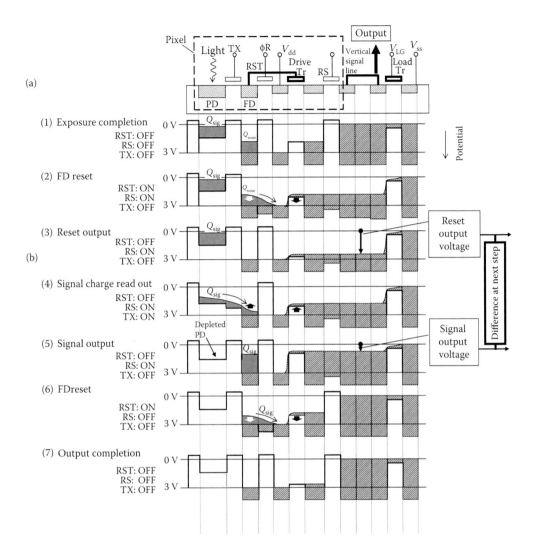

FIGURE 5.47
Operational diagram of one pixel in a 4-Tr configuration: (a) cross-sectional view; (b) operational diagram by potential distribution.

of metal wiring is necessary with the *n*-type region of the PD in a 4-Tr pixel, it is possible to cover the *n*-type region with a p^+ layer to form a buried PD to suppress dark current drastically. As capacitance of FD can be designed without reference to PD performance, it is possible to realize a higher charge–voltage conversion factor, which is inversely proportional to C_{in}. These are huge advantages of 4-Tr pixel over 3-Tr pixel CMOS sensors. Therefore, CMOS sensors aiming for high performance employ a 4-Tr pixel configuration.

The noises in 4-Tr pixel configuration are reset noise of FD, FPN caused by variation of threshold voltage, 1/f noise, and thermal noise of the drive transistor in SFA. Among these, reset noise of FD, FPN due to V_{th} variation, and part of 1/f noise of the drive transistor can be removed by a cancellation operation by CDS. The high-frequency component of thermal noise is removed due to the narrow bandwidth of CDS.

5.3.2.3 Shared Pixel Architecture

A pixel of a IT-CCD is made up of one PD and two electrodes of a CCD, as shown Figure 5.10. While the PD is surrounded by isolation, because each part can be arranged without a separator, a pixel is formed in a small area. On the other hand, one PD and four transistors are necessary for CMOS sensors. One transistor is made up of one gate electrode, two diffusions for the source and drain, and a separator surrounding them. A larger area is necessary to form 4-Tr pixel configuration. This is the main reason that CMOS sensors were not produced in quantity until the late 1990s, while MOS sensor and CCD production started in the early 1980s. Small size MOSFET technology, with CMOS processing technology around 0.35 μm, which enabled 4-Tr pixel configuration, had to be realized first to lessen the cost restriction. Still, a 4-Tr pixel needs four transistors.

As a method of pixel shrinkage, shared pixel architecture,[34] in which one FDA is shared by plural pixels, is mainstream in the field of small pixel sensors. In an example of two-shared pixel configuration shown in Figure 5.48a, reset transistor (RS), drive transistor (amplify transistor), and row selection transistor (RS) are shared by two pixels. The elements of each pixel that are not shared are the PD, transfer (readout) transistor (TX), and FD, and FDs are connected by wiring. Since there are five transistors for two pixels, the number of transistors per one pixel is decreased to 2.5, that is, 2.5 Tr/pixel. An example of a four-shared pixel is shown in Figure 5.48b. As the shared parts are the same as in Figure 5.48a, seven transistors exist for four pixels, that is, 1.75 Tr/pixel. An increase in the number of shared pixel causes capacitance increase of FD, since the number of

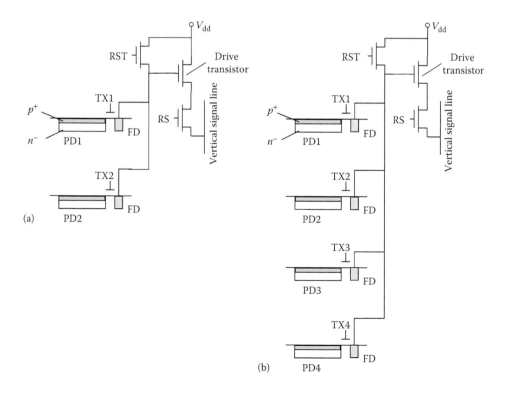

FIGURE 5.48
Shared pixel configuration: (a) two-shared pixel configuration (2.5 Tr/pixel); (b) four-shared pixel configuration (1.75 Tr/pixel).

connected FDs increases. This means a decrease in the charge–voltage conversion factor, that is, decrease of output voltage and charge quantity gain. Therefore, a balanced design is required for overall performance.

A means to reduce transistor number further is FD-drive architecture,[35,36] illustrated in Figure 5.49. Figure 5.49a and b show examples of vertical four-shared pixel configuration and horizontal and vertical four-shared pixel configuration, respectively. In Figure 5.49a, the row select transistor (RS) is removed in comparison with Figure 5.48b.

In this type, FD potentials are kept lower than those in a situation filled with charges are. At the row select operation, only the FDs of pixels belonging to the selected row are set to the high voltage condition of clock voltage source V_{dd} by way of RST. Then, signal charges are transferred from PD to FD as well as a normal 4-Tr pixel sensor. By this setting, gate electrodes connected to FDs of driver transistors are set to lower voltage than that of the selected pixel. Let us consider a situation where a number of drive transistors are commonly connected to a vertical signal line in (5) in Figure 5.47, and among them only one drive transistor has a gate electrode with higher voltage than the others. It is clear that electrons flowing from V_{ss} to V_{dd} take the channel of lowest potential energy determined by the selected pixel. This means that drive transistors of unselected rows are the same in the off-situation, that is, it is equal to a situation where they are not selected, namely, winner takes all.

In this configuration, six transistors are used for four pixels, so the number of transistors per pixel is 1.5 Tr/pixel. In the case of a horizontal and vertical shared pixel, since FDs are also shared by two horizontal pixels, as shown in the Figure 5.49b, this type has the advantage of lower FD capacitance, that is, suppression of charge–voltage conversion factor decrease.

Since vertical shared pixel configuration has translation symmetry in both horizontal and vertical directions, misalignment at patterning process gives the same effect to all pixels. Performance uniformity between pixels is kept. On the other hand, horizontal and vertical shared pixel configuration has translation symmetry only in the vertical direction.

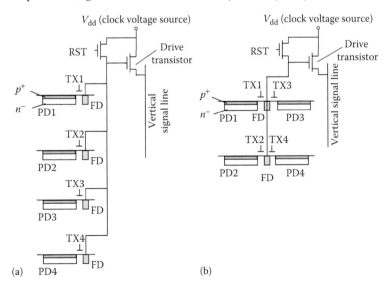

FIGURE 5.49
FD-drive shared pixel configuration: (a) vertical four-shared pixel configuration (1.5 Tr/pixel); (b) horizontal and vertical four-shared pixel configuration (1.5 Tr/pixel).

Since it has reflection symmetry in the horizontal direction, horizontal misalignments are apt to give opposite effects to left-side and right-side pixels, that is, performances differ between even and odd columns. There is an example of an eight-shared pixel sensor[37] by sharing two pixels horizontally and four pixels vertically.

Movements of charges in shared pixel architecture are realized by connecting individual parts by metal wiring. This is an effectual measure for pixel shrinkage for CMOS sensors that CCDs cannot use. On the other hand, it might be possible to recognize the CCD as the ultimate shared pixel sensor, because only one FDA is shared by all pixels through noiseless connections.

5.3.3 Progress in CMOS Sensors

As just described, pixel configuration and device constitution of CMOS sensors have converged, pursuing an improvement in performance level to compete against CCDs. More specifically, a 4-Tr pixel configuration with a buried/pinned PD and column CDS circuit has been adopted. By reducing pixel noise in this manner, it becomes more important to suppress noise of the following read circuits. An inevitable approach for image sensors was to improve SNR by noise reduction in the analog domain circuit as the first step. Subsequently, along with noise reduction and speeding up through digitization of light intensity signal, sensitivity improvements by making effective use of incident light are being approached.

5.3.3.1 Noise Reduction Circuits in Analog Output Sensor

Signal amplification forms the basis of reduction of impact of noise of the following circuits. Therefore, newly added FPN caused by new amplification is removed by additional FPN cancelation circuits.

5.3.3.1.1 Amplifying Noise Canceller

Since the configuration is an analog circuit, charge quantity multiplied by pixel amplifier drives the following circuit as the input charge quantity at the input part. Although it is desirable to amplify as much as possible in the earlier stage before new noises are added, this often fails because of the limitations of the available voltage range and capacitance volume. A proposed method to reduce the impact of noise after a CDS circuit is column amplifier configuration. While it is possible to arrange an amplifier before or after the CDS circuit in Figure 5.46, the proposal is a circuit in which the vertical signal line is followed by a column amplifier and there is an offset variation canceling function,[38] shown in Figure 5.50.

When FD is reset, switches ϕ_1 and ϕ_2 in the column amplifying noise canceller are made on-state to set the input part of the column amplifier as clamped. After ϕ_2 is changed to off-situation, the readout pulse is applied to TX to transfer signal charges in PD to FD, and amplified difference signal is output by the column amplifier. The voltage gain G of the column amplifier is given by $G = C_1/C_2$. The higher the gain is, the higher the noise suppression effect is. But this is meaningless if the amplified signal saturates the voltage range of the circuit. Moreover, higher gain is necessary for a low-level signal itself to obtain a higher SNR. Therefore, the column amplifier selects gain adaptively by choosing the best capacitance as C_2.

In this configuration, noise of the column amplifier itself, pixel amplifier, and noise canceller are suppressed as well as reduction effects of noise impact of substantial-stage

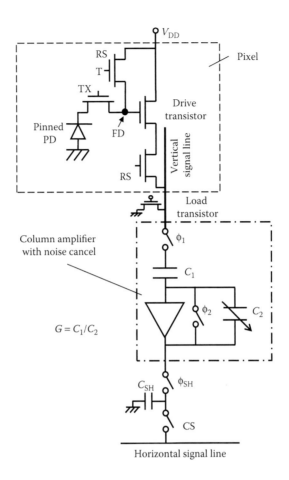

FIGURE 5.50
Column amplification with noise canceling function. (Reprinted with permission from Kawahito, S., Sakakibara, M., Handoko, D., Nakamura, N., Satoh, H., Higashi, M., Mabuchi, K., and Sumi, H., *Proceedings of the IEEE International Solid-State Circuits Conference, Digest of Technical Papers*, 12.7, pp. 224–225, 2003.)

circuits to $1/G$. As the column amplifiers work in parallel in bandwidths <1 MHz, noises in higher frequency ranges generated in anterior stages, such as thermal noise in the transistors in the SFA, are also reduced. Amplified difference signal is sampled and held in sampling capacitance C_{SH} at each column and output to the horizontal signal line by way of the column select transistor CS.

5.3.3.1.2 High-Gain Double-Stage Noise Canceller

By amplifying the noise canceller described above, not only noise of both pixel and column amplifiers, but also noise caused by the noise canceller are removed. The impact of noise in the noise canceller is also suppressed. If column amplifier gain is high enough, the only remaining noise is that originating from variation of charge quantity at the input node of the column amplifier caused by kTC noise, which occurs when ϕ_2 is changed to off-state. In a proposed configuration,[39] the amplifying noise canceller in Figure 5.50 is followed by the differential circuit discussed in Figure 6.42a as the second-stage differential circuit, as shown in Figure 5.51.

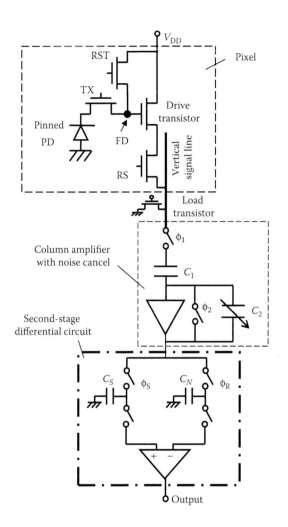

FIGURE 5.51

High-gain double-stage noise canceller. (Reprinted with permission from Takahashi, H., Noda, T., Matsuda, T., Watanabe, T., Shinohara, M., Endo, T., Takimoto, S. et al., *Proceedings of the IEEE International Solid-State Circuits Conference, Digest of Technical Papers*, 28.6, pp. 510–511, San Francisco, CA, 2007.)

Outputs of the column amplifier corresponding to reset and signal level situations are sampled and held in C_N and C_S, respectively, and their difference is output by the second-stage differential circuit. As mentioned in Figure 5.42a, this type of differential circuit is not formed at each column but arranged as a common circuit at the output part, and it needs a wide bandwidth.

5.3.3.2 Digital Output Sensors

As described in Sections 1.2 and 1.3, three (space, wavelength, time) of the four factors of image information are already digitized as built-in coordinate points in imaging systems. Image sensors only detect light intensity coming into each coordinate point. Currently, signal voltage proportional to light intensity is obtained by detecting potential change of FD, to which signal charges are transferred. Thus, light intensity has been treated as an analog quantity in this book so far.

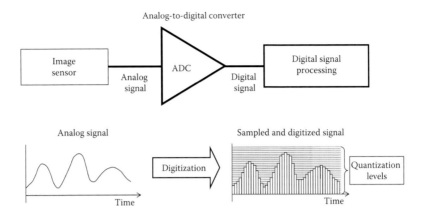

FIGURE 5.52
Digitization of signals.

Analog output from sensors was processed* and recorded to videotapes as an analog voltage signal until the mid-1990s when digital video cameras appeared on the market. But digital signal systems were strongly required from viewpoints of interoperability (it is easier to expand to different systems), media, and lossless copies. Furthermore, since a digital signal is superior to analog processing with respect to variety functions and stability, analog sensor output was digitized by an analog-to-digital converter (ADC) before signal processing, as shown in Figure 5.52.

This is desirable from an SNR viewpoint. Because analog can take any value, noise can be added to analog signals. Digital signals are allowed to take only one of the quantized values[†] in the system. Thus, digital signal systems are tolerant to noise addition, which is also an advantage as no new noise occurs after conversion to digital signal.[‡] In that sense, digitization at an earlier stage is favorable. Some new functions such as high-speed readout described below can be added, depending on the digitization stage and method.

Sampling and quantization are necessary for analog-to-digital (A/D) conversion as shown in the bottom right of Figure 5.52. Sampling means to sample analog signal in a specific frequency. In "L" bit digital systems, the discrete value of only the number of 2^L is allowed to be taken. Quantization means to replace the sampled analog value to the nearest discrete value in 2^L values. There are three stages at which light intensity information is digitized, as follows:[40,41]

1. Chip-level ADC sensor. This is a configuration in which independent commonly used ADC devices are formed on the same chip with sensors, as shown in Figure 5.53a. No heavy development subject is necessary. This type needs analog output of sensor and ADC having the same range of frequency (about 30 MHz) with a case of independent chip configuration. There is no function or bandwidth advantage, while high-frequency analog output from a chip is not necessary.

2. Column-level ADC sensor. This type has been widely put to practical use in consumer applications. An ADC converter is formed at each column, as shown in

---

* There was a time when analog signal was converted to digital signal to apply highly functional digital signal processing, and then converted back to analog signal for analog recording.
† Of course adequate resolution for the specific application is necessary.
‡ Although it is possible for the noise component rate to increase by arithmetic processing, this does not mean new noise is added.

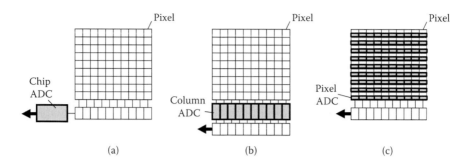

FIGURE 5.53
Stages of signal digitization: (a) chip-level ADC; (b) column-level ADC; (c) pixel-level ADC.

Figure 5.53b. Since A/D conversion is parallelized in lower bandwidths, <1 MHz, it has an advantage in noise performance. Digitized signal can be output at speeds 10 times higher than in chip-level ADC sensors.

3. Pixel-level ADC sensor. Stretching much further back to the source, an ADC is formed at each pixel in this configuration, as shown in Figure 5.53c. Because of pixel-parallel processing, A/D conversion in even lower bandwidths is possible. Currently these sensors are still being researched[42] and are for special use only, because many transistors are necessary at each pixel, among other reasons.

5.3.3.2.1 Chip-Level ADC Sensor

An example of a chip-level ADC sensor is shown in Figure 5.54.[43] Since an analog signal with pixel output frequency around 30 MHz is digitized, the same range of bandwidth of ADC is necessary. Green pixel signals and red/blue pixel signals are read out from the top and bottom channels, respectively, making this sensor dual channel. Each channel has a pipeline ADC, which is suitable for high-speed conversion by parallel processing serial signals. Since circuit size increases in proportion to step number, pipeline ADCs are

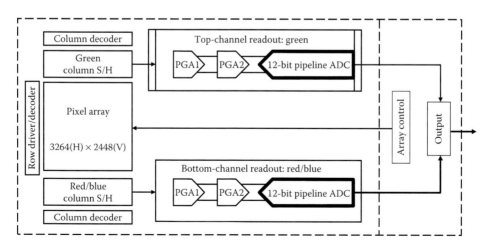

FIGURE 5.54
Block diagram of a chip-level ADC sensor. (Reprinted with permission from Cho, K., Lee, C., Eikedal, S., Baum, A., Jiang, J., Xul, C., Fan, X., and Kauffman, R., *Proceedings of the IEEE International Solid-State Circuits Conference, Digest of Technical Papers*, 28.5, pp. 508–509, 2007.)

hard to use as column-level ADCs. But they are suitable for chip-level ADCs, because they have the advantage of high-speed performance. This sensor has 1.75 Tr/pixel in a four-shared pixel configuration with 1.75 µm pixel pitch and total pixel number of 8.1 M. The maximum frame rate is 12 frames per second (fps) with a 12-bit output.

5.3.3.2.2 Column-Level ADC Sensor

Although FPN originating from pixel characteristics can be removed by an offset cancellation circuit formed at each column, light vertical bars caused by performance variation of cancelation circuits themselves remain. And as a certain volume of capacity is necessary to contain analog signal in the CDS circuit, the column circuit becomes vertically longer for capacitor size along with pixel pitch shrinkage. A column ADC is an effective step for this subject. As column-level ADCs, the single-slope integrating type in which time information is used, successive-approximation type and cyclic type in which digital code is encoded from the most significant bit, and ΔΣ type have been proposed.

5.3.3.2.2.1 Single-Slope Integrating Type ADC The column-level ADC type that is used most often in practice is the single-slope integrating type.[44,45] By combination clumping in the analog domain with CDS in the digital region, a very low FPN level is achieved. As Figure 5.55 shows, it is made up of the following six parts: (1) pixel array, (2) row decoder, (3) column-parallel ADC, (4) digital-to-analog convertor (DAC) to generate ramp wave, (5) logic control circuit, (6) and digital output interface. This sensor is driven with a 75 MHz master clock. By the generation of four times the 300 MHz frequency of a phase-locked loop circuit, the column-parallel ADC, single-slope ADC, and low-voltage differential

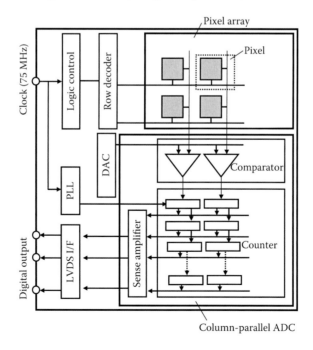

FIGURE 5.55
System block diagram of a single-slope integration-type ADC sensor. (Adapted from Yoshihara, S., Kikuchi, M., Ito, Y., Inada, Y., Kuramochi, S., Wakabayashi, H., Okano, M., et al., *Proceedings of the IEEE International Solid-State Circuits Conference, Digest of Technical Papers*, 27.1, pp. 1984–1993, San Francisco, CA, 2006. With permission.)

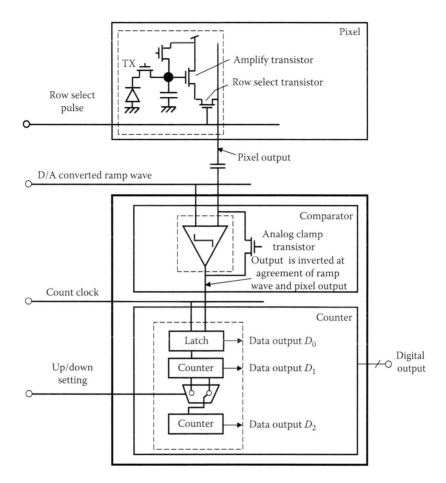

FIGURE 5.56
Block diagram of a single-slope integration ADC. (Reprinted with permission from Nitta, Y., Muramatsu, Y., Amano, K., Toyama, T., Yamamoto, J., Mishina, K., Suzuki, A., et al., *Proceedings of the IEEE International Solid-State Circuits Conference, Digest of Technical Papers*, 27.5, pp. 2024–2031, San Francisco, CA, 2006.)

signaling interface are processed at high speed. Pixel output is connected to comparator input by way of coupling capacitance, and ramp wave is applied to another input, as shown in Figure 5.56 in more detail.

Comparator output is followed by a counter to which an up/down setting signal to specify count direction is input to achieve CDS processing in the digital domain. In operation, at a stage of reset of FD in the pixel, pixel signal input and output of the comparator is connected by an analog clamp transistor to the clamp reset signal level. Since this operation is the same as the clamp in analog CDS, variation of reset level as input signal to comparator is removed. This is also effective to shorten the reset signal count period, as will be described later.

The clock number from the PLL circuit is counted to convert the reset signal to digital code until the comparator output is inverted by agreement of ramp wave and pixel output. Ripple counters are set to down count by up/down setting signal during reset signal and are counted as shown in the timing chart in Figure 5.57.

After that, signal charges in the PD are read out to FD to output sensor signal at pixel output. They are counted in the same manner with reset signal except for being set to up counting. Since the counting directions of reset and sensor signals are opposite, the

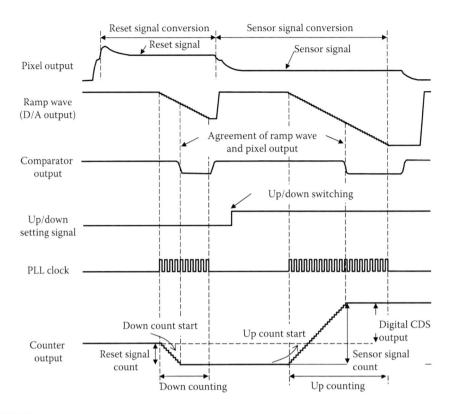

FIGURE 5.57
Timing chart of digital CDS by single-slope integration. (Reprinted with permission from Nitta, Y., Muramatsu, Y., Amano, K., Toyama, T., Yamamoto, J., Mishina, K., Suzuki, A., et al., *Proceedings of the IEEE International Solid-State Circuits Conference, Digest of Technical Papers*, 27.5, pp. 2024–2031, San Francisco, CA, 2006.)

counter digitally subtracts conversion of the reset signal from the sensor signal, as shown by counter output in Figure 5.57. In this way, the difference between the sensor and reset signals is obtained in the digital domain, so it is called digital CDS. The obtained digital data are transferred to data latches in each counter block. The horizontal data transfer and A/D conversion of the next row are processed in parallel.

This configuration has the advantage of low-noise performance because of not only highly accurate offset noise cancelation but also a lower band of frequency, thanks to A/D conversion by column-parallel processing. By column-parallel A/D conversion, the sensor achieves a high frame rate of 180 fps[44] in spite of the larger pixel number of 2.8 M. The digital resolution is 12 bits. The noise of the electron number of column FPN by digital CDS is reported to be 0.5.

As circuits of single-slope integrating type ADC are rather simple, they can be used for narrow-pitch pixel sensors. They are suitable for image sensors because of their lower power consumption. On the other hand, a significantly shorter conversion time is necessary to increase the number of bits for digital coding. It is not easy to achieve both high speed and high digital resolution.

5.3.3.2.2.2 Successive-Approximation-Type ADC   Successive-approximation ADC (SA-ADC) is used as column ADC to realize high-speed conversion. Digital codes of all bits are counted, starting from the most significant bit and finishing at the least significant bit. DAC

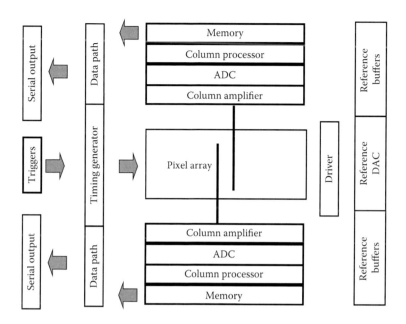

FIGURE 5.58
Block diagram of a 14-bit successive-approximation-type ADC sensor. (Reprinted with permission from Matsuo, S., Bales, T.J., Shoda, M., Osawa, S., Kawamura, K., Andersson, A., Haque, M., et al., *Transactions on Electron Devices*, 56(11), 2380–2389, 2009.)

is equipped to reconvert once-converted digital code to analog signal to compare with analog signal input by the comparator to judge the digital code of the second most significant bit. By repeating this comparison, the precision of digital code is raised one bit at a time. Although SA-ADC is operated N times if the bit number is N, A/D conversion is completed by N-times repetition using only one comparator. While repletion time is necessary, it can be designed in a compact circuit size.

Figure 5.58 is a block diagram of a 14-bit SA-ADC sensor.[46] Around the pixel array, the control part and output are arranged on the left side, the driver and DAC are arranged on the right side, and the column ADCs are arranged on the top and bottom.

Figure 5.59 is a circuit diagram of an SA-ADC. Pixel output is amplified by a variable gain amplifier for noise reduction. An ADC consists of a dual comparator and DAC. In DACs, capacitors are arranged with the necessary numbers and volumes to deal with multiple bits. Three different reference voltages, consisting of standard voltage and 1/4 and 1/16 of the standard voltage, are used to suppress increase of the circuit size by decreasing the number of capacitance volumes. While SA-ADC generally excels in high-speed performance and low consumption, the ratio of area size that capacitances occupy increases for higher digital resolution.

This sensor has a pixel pitch of 4.2 μm, and a higher frame frequency of 60 fps for the large pixel number of 8.9 M. The number of column FPN electrons is suppressed to 0.36 owing to digital CDS.

5.3.3.2.2.3 Cyclic-Type ADC In cyclic ADC, digits are counted starting from the most significant bit by one-bit-at-a-time comparison as well as successive-approximation type. Since resolution is determined by cyclic number differently from in SA-ADC sensors, it has the advantage of compatibility between high-speed performance and high resolution.

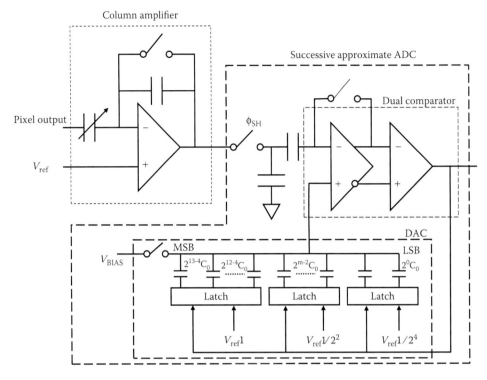

FIGURE 5.59
Circuit diagram of a 14-bit successive-approximation ADC. (Adapted from Matsuo, S., Bales, T.J., Shoda, M., Osawa, S., Kawamura, K., Andersson, A., Haque, M., et al., *Transactions on Electron Devices*, 56(11), 2380–2389, 2009. With permission.)

Figure 5.60 is a block diagram of a cyclic-type ADC sensor.[47] Although circuit configurations were complicated in origin, an architecture applicable to image sensors was proposed.[48] A circuit configuration of a 13-bit cyclic column-parallel ADC is shown in Figure 5.61.

This ADC is made up of an operational transconductance amplifier (OTA), capacitance ($C_1 = C$, $C_2 = C/2$), sub-ADC composed of two comparators, and DAC. Before A/D conversion, the system is initialized by setting all the switches on-state to reset all charges remaining in the capacitances. This operation enables high-precision digital CDS. During sampling, inputs are applied to the sub-ADC to determine the most significant bit, and the sub-ADC compares V_{IN} with two reference voltages (V_{RCH}, V_{RCL}) and generates three kinds of digital code D ($[D_1, D_0]$), whose values are 0, 1, or 2. Switches in the DAC are controlled by using the digit of the most significant bit. As capacitances (C_{1a} and C_{2b}) that are connected to the inverting input terminal of the amplifier are connected with V_{RCH} or V_{RCL}, the input V_{IN} is multiplied by two and a reference level is subtracted from it. Then, amplifier output is sampled by capacitances C_{1a} and C_{1b} and fed back for the second significant bit determination. Multiplications and feedbacks are repeated until the necessary digit codes are obtained. To obtain 13 bits, they are repeated 11 times. This sensor has 5.6 μm pitch and 300 k pixels with very-low-level column FPN of 0.1 electrons and a high frame rate of 390 fps.

5.3.3.2.2.4 ΔΣ-Type ADC   Figure 5.62 is a block diagram of an ADC sensor[49,50] developed for realization of low noise and wide dynamic range by using a ΔΣ ADC with a narrow circuit area and low power consumption. It is generally known that ΔΣ ADCs achieve high

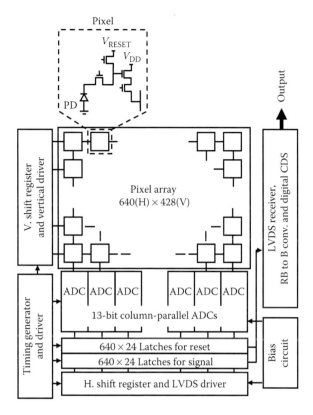

FIGURE 5.60

Block diagram of a 13-bit cyclic-type ADC sensor. (Reprinted with permission from Park, J., Aoyama, S., Watanabe, T., Isobe, K., and Kawahito, S., *Transactions on Electron Devices*, 56, 2414–2422, 2009.)

resolution by oversampling and noise shaping through a highly precise analog circuit. The sensor is made up of a pixel array, column-parallel $\Delta\Sigma$ ADC, SRAM buffer memory, and controller. An ADC consists of a $\Delta\Sigma$ modulator and decimation filter. The device has dual-channel output configuration with top and bottom outputs and a two-shared 2.25 μm pitch with 2.1 M pixels. Signal from the pixel is digitized by a $\Delta\Sigma$ ADC and transferred to SRAM, and output by way of a sense amplifier during the next row is converted.

Figures 5.63 and 5.64 are schematic diagrams of the column readout circuit and timing diagram, respectively, of one horizontal conversion. A high speed of 120 fps is realized with a horizontal period of 6.85 μs. In operation, pixel reset level output is input to a second-order $\Delta\Sigma$ modulator. Since the $\Delta\Sigma$ modulator oversamples (110 times in this case) input signal during conversion, there is a noise reduction effect by averaging. As modulator output including noise is filtered by a second-order decimation filter whose output increases as a second-order function of time, the number of clocks necessary to represent digital value is considerably less than that in the cyclic type. It is therefore possible to realize high-speed conversion. For one A/D conversion, pixel output is sampled 110 times by a 48 MHz clock in 2.3 μs. After reset level conversion is completed, the clock stops and each bit digit of the reset level is inverted. That is, $\overline{D_{RST}}$ is obtained from D_{RST} as shown in the bottom of Figure 5.64. This negative reset value is used as the initial value of the second conversion for the signal level of the pixel. Since the difference of the digital value, which is reset level subtracted from signal level, is obtained by this operation, digital CDS is realized.

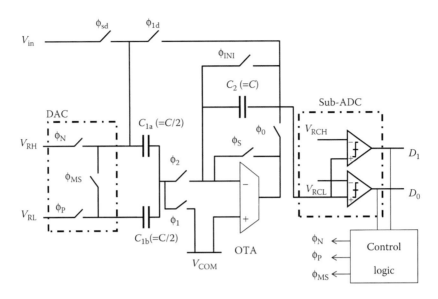

FIGURE 5.61
Circuit configuration of a 13-bit cyclic column-parallel ADC. (Reprinted with permission from Park, J., Aoyama, S., Watanabe, T., Isobe, K., and Kawahito, S., *Transactions on Electron Devices*, 56, 2414–2422, 2009.)

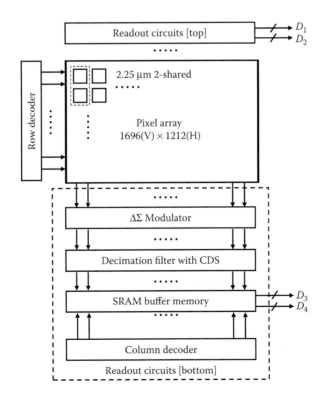

FIGURE 5.62
Block diagram of a 12-bit ΔΣ ADC sensor. (Reprinted with permission from Chae, Y., Cheon, J., Lim, S., Lee, D., Kwon, M., Yoo, K., Jung, W., Lee, D., Ham, S., and Han, G., *Proceedings of the IEEE International Solid-State Circuits Conference, Digest of Technical Papers*, 22.1, pp. 394–395, San Francisco, CA, 2010.)

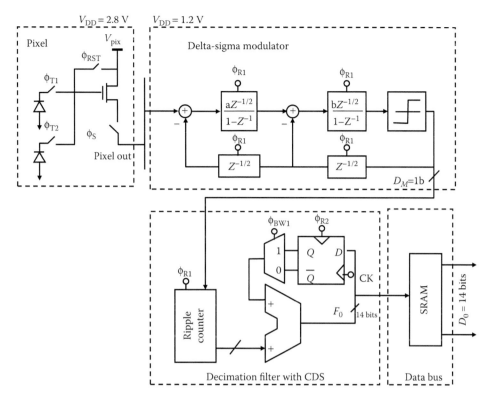

FIGURE 5.63
Schematic diagram of the column readout circuit. (Reprinted with permission from Chae, Y., Cheon, J., Lim, S., Lee, D., Kwon, M., Yoo, K., Jung, W., Lee, D., Ham, S., and Han, G., *Proceedings of the IEEE International Solid-State Circuits Conference, Digest of Technical Papers*, 22.1, pp. 394–395, San Francisco, CA, 2010.)

The reasons second-order $\Delta\Sigma$ ADCs are adopted rather than first-order ones are as follows:

1. Conversion 2^N times is necessary to obtain N bits in the case of a first-order $\Delta\Sigma$ ADC, and a high-speed clock over 1 GHz is necessary to obtain 12 bits.
2. Since pixel signal to be converted is DC signal, the number of times of conversion is limited and it is not converted accurately by a first-order $\Delta\Sigma$ ADC.
3. In the case of first-order $\Delta\Sigma$ ADCs, high amplification gain of about 70 dB is required to avoid band of frequency loss at conversion.

Thus, it is considered that this sensor needs to elaborate the well-balanced design in the 2.25 μm pixel pitch with dual-channel output by adopting a second-order $\Delta\Sigma$ ADC to achieve high performance, such as 12-bit resolution at a high frame rate of 120 fps, 13-bit resolution at 60 fps, low-noise performance, low power consumption, and a high dynamic range of the ADC.

As described above, column-parallel ADC sensors have been realized by various types of ADC, and some have become a commercial reality. Depending on the development of CMOS process technology such as downscaling and circuit design technology, the boundary conditions for the best technical choice might change. This should be aimed for in any future development.

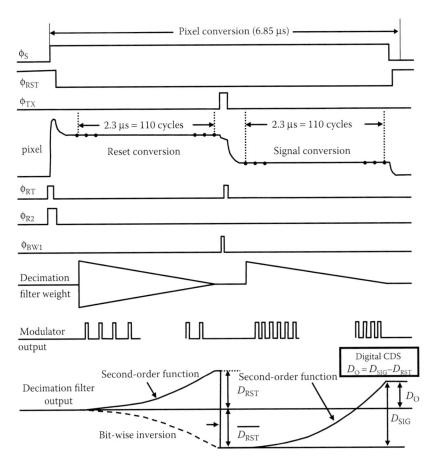

FIGURE 5.64
Timing diagram of one horizontal conversion. (Reprinted with permission from Chae, Y., Cheon, J., Lim, S., Kwon, M., Yoo, K., Jung, W., Lee, D., Ham, S., and Han, G., *Journal of Solid State Circuits*, 46, 236–247, 2011.)

5.3.3.2.3 Pixel-Level ADC Sensor

If A/D conversion of light intensity information is done in the pixel, no new noise is added through the readout pass. This is not only a great advantage in terms of SNR, but also a very important functionality because of flexible operation or signal processing at intra- and interpixel levels. As processing is at each pixel level, it is the ultimate parallel treatment and also benefits high-speed performance. Depending on the method and application, the frame-based accumulation mode, which almost all image sensors adopt, might not be necessary. It is highly expected that flexibility and high functionality in the time domain will be improved spectacularly.

While there are examples of all kinds of ways to develop pixel-level ADC sensors through the front door, a unique sensor is introduced here. Although this sensor is usually categorized as a wide dynamic range sensor, it can be considered as a pixel-level ADC sensor. But no so-called "A/D converter" circuit is used. This sensor is the first example that does not belong to "almost all image sensors" in this book, as described in the first footnote in Section 1.2.3. The pixel circuit diagram and output of pulse photosensor[51] are shown in Figure 5.65a and b, respectively.

The n-type region of the PD is connected to voltage source V_{dd}, and the reset transistor is formed between the p-type region of the PD and GND level. The p-type region is connected to input of the first stage of an integrated four-stage CMOS inverter chain. The voltage of the p-type region, which is at GND level just after reset operation, rises along with an increase in signal charges (hole). When the voltage reaches the threshold voltage of the first-stage inverter, the inverter output is inverted, which causes inverts of the following inverters. The output of the final-stage inverter turns to high or "1" from low or "0." It makes reset transistor switch on-state to reset the p-type region of the PD to GND level. Then, inverts of outputs of four inverters follow in series. Consequently, final-stage inverter output is switched to low or "0," which makes the reset transistor off-state to complete the reset operation and the next signal charge integration starts. By a series of operations, one pulse is output from the final stage of inverter chains. By using the PD capacitance as one unit to measure charge quantity, one pulse is output and the PD is reset when a predetermined charge quantity is integrated in the PD. Since higher-frequency clock pulse is output for higher illumination, while the opposite situation occurs for darker areas, as shown in Figure 5.65b, it can be considered that light intensity information is converted to the frequency of output pulse. Only by counting the pulse number to code digitally is pixel-level A/D conversion achieved. The reason a four-stage inverter chain is adopted is to make time for the reset transistor to reset the operation with certainty by time delay of the inverters.

In this sensor, quantized factors are space and amount of change of integrated charge quantity in PD. So the signal output of the sensor is not integrated signal charge quantity S,

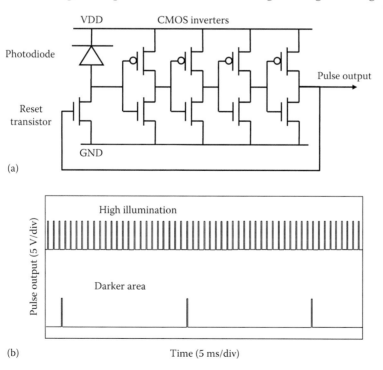

FIGURE 5.65
Pulse output sensor: (a) pixel circuit diagram; (b) pulse output. (Reprinted with permission from Yang, W., *Proceedings of the IEEE International Solid-State Circuits Conference, 41st ISSCC Digest of Technical Papers*, 13.7, pp. 230–231, San Francisco, CA, 1994.)

but time T when amount of integrated signal charge quantity reaches a predetermined amount ΔS_q at pixel r_k, that is, $T(\Delta S_q, r_k)$.

Because processing circuits are necessary at each pixel in pixel-level ADC sensors and therefore many transistors are required for each pixel (among other problems), they are still under investigation and are only used for special purposes. But if it was possible to use them widely, noise in the readout operation would not be a problem and they would be very easy devices to deal with. This prediction is reminiscent of the way that CMOS sensors have become accepted and mainstream as a result of progress such as transistor shrinkage despite needing multiple transistors and FDAs in each pixel, while CCDs need only one common FDA.

At the advanced phase of realization of pixel-level ADC sensors, no noise except optical shot noise would not be allowed. It seems doubtful to think that a solution of the pixel-level ADC sensor problem would be an extension of the shrunken version of the present ADC circuit.

In photoelectric conversion, one photon is absorbed and one signal charge is generated. Since a photon is a quantized particle of light, the most precise measurement of light intensity should be to count photons one by one, if possible. Although there are methods of counting photons, they are available only in situations where photons come with enough temporal intervals, that is, very-low-level illuminance in the present situation. These methods cannot be applied to high-illumination situations, in which many photons arrive to a sensor part concurrently or continuously. If it were possible to count each photon directly, this would be an ideal solution. It is hoped that new concepts and technology will emerge.

5.3.3.3 Sensitivity Improvement Technology

Sensitivity in terms of SNR is the most important and eternal performance issue for image sensors. While noise reduction approaches were discussed in Sections 5.3.3.1 and 5.3.3.2, signal-increasing approaches try to decrease optical loss using techniques such as reflection, absorption, and cross talk through paths to the PD.

5.3.3.3.1 Lightguide (Lightpipe)

On-chip micro lenses (OCLs), discussed in Section 5.1.2, have been very successful as a means to leading incident light to the sensing aperture, especially in CCDs. In the early 1980s, OCLs were proposed for situations where pixel pitches were around 10 µm with an aperture of about 5 µm, about ten times wider than the present pitch and aperture. Focusing at the aperture means sharpshooting it by incoming light to focus on it. But the incident light angle includes not only the perpendicular component but also a lot of the oblique component, which decreases the efficiency of the light focus. An inner lens was developed to suppress degradation of light focus efficiency caused along with aperture shrinkage and shortening of lens focusing length.

In CMOS sensors, light focus efficiency decreases more because the distance from the top portion of the OCL to the PD is lengthened by applying plural metal layers for wiring as well as CMOS logic LSIs. Although three-dimensional wiring by multiple metal layers enables smaller chips for cost reduction in logic devices, it inhibits optical performance such as lower sensitivity and higher cross talk. A lightpipe was developed to suppress this type of degradation. Especially in fine-pitch pixels, as optical path widths are close to light wavelength, it is not easy to arrive at a corresponding PD for incident light because of shielding, reflection, and diffraction caused by metal wiring through the paths shown in Figure 5.66a.

Some incident light reaches the next PD and this is *cross talk*. In lightpipe sensors, the optical paths between the color filter and photodiode are filled with high refractive index

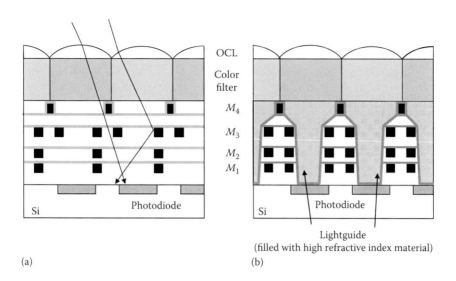

FIGURE 5.66
Schematic diagram of a CMOS sensor with a four-metal-layer wiring pixel: (a) without a lightguide; (b) with a lightguide. (Reprinted with permission from Gambino, J., Leidy, B., Adkisson, J., Jaffe, M., Rassel, R.J., Wynne, J., Ellis-Monaghan, J. et al., *Proceedings of the International Electron Devices Meeting, Technical Digest*, pp. 5.5.1–5.5.4, San Francisco, CA, 2006.)

material,[52] as shown in Figure 5.66b. While OCLs correspond to sharpshooting, as it were, in lightpipe sensors incident light is confined in high refractive index material and guided to the PD, similar to a funnel. Using this method, loss of light and cross talk can be suppressed.

5.3.3.3.2 Backside Illuminated Sensors

Backside illuminated sensors (BSIs) have a structure in which incident light can directly reach the PD without passing through metal layers by receiving light from the backside or the substrate side of sensors. Although BSIs were proposed with the first FT-CCD[4] in 1972, processing to fabricate thin silicon wafers is complicated and expensive. This was adopted only for special applications such as scientific measurements[53] and was used only for monochrome image sensors. When a BSI-type CMOS sensor[54] was proposed for a consumer SCCC, a big trend[55] was started. Schematic diagrams of cross-sectional views of a front-side illuminated sensor (FSI) and a BSI are compared in Figure 5.67.

While incident light has to pass through metal wiring layers in the FSI, light can arrive at the PD directly in the BSI. Therefore, the BSI has some advantage in sensitivity over the FSI. This is very easy to understand through instinct, that is, advantageous technology for monochrome sensors. For this reason, not only people who are unfamiliar with image sensor technology but also those who should be experts on it often misunderstand that the BSI has an absolute advantage in sensitivity over the FSI without any conditions. This will be described in detail in this section and Section 5.3.3.3.3.

From the viewpoint of pixel design, flexibility of metal layout is of great benefit in BSIs.

The process flow shown in Figure 5.68 is necessary to implement a BSI structure. (a) At the first PD, a circuit and wiring are formed, as well as an FSI. (b) Then, a supporting substrate is attached to the surface side. (c) Wafer thinning is followed by grinding and etching. (d) New backside surface is processed to suppress dark current and a light shield film is formed. (e) Formation of color filters and OCLs completes the process flow. The steps

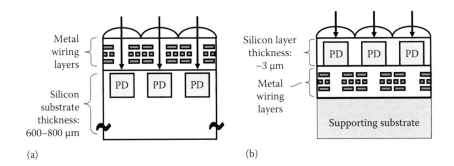

(a)   (b)

FIGURE 5.67
Schematic diagram of a simplified cross-sectional view of (a) a surface illustrated sensor (FSI); (b) a backside illuminated sensor (BSI).

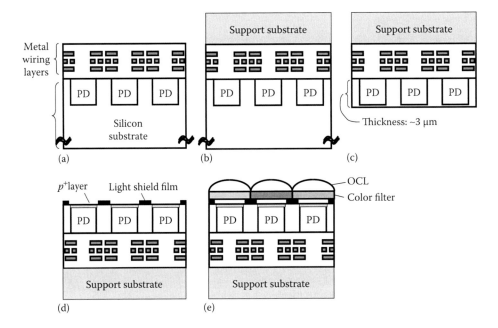

FIGURE 5.68
Conceptual diagram of the fabrication process of a BSI: (a) formation of circuitry and wiring on a silicon wafer; (b) attachment of support substrate; (c) wafer thinning; (d) backside passivation and light shield formation; (e) color filter and OCL formation.

shown in Figure 5.68b–d were newly added for BSIs. Image sensors are a kind of LSI and have been naturally manufactured by technology cultivated in the LSI industry including the fabrication process. Technologies inherent in image sensors hitherto include intrinsic gettering (IG),[56] IG-epi wafers,[57] buried/pinned PDs, color filters, OCLs, glass-sealed packages, and so on. BSIs have a greatly expanded technological area endemic to image sensors quantitatively and qualitatively. While newly added processes increase the manufacturing cost, they should bring about benefits that outweigh this cost increase for the industry.

Figures 5.67 and 5.68 show conceptual drawings. High optical performances, such as sensitivity, are expected of BSIs, especially in smaller size pixels.

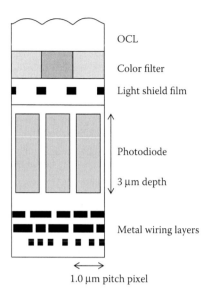

FIGURE 5.69
Cross-sectional view showing the aspect ratio of a substantial BSI of 1.0 μm pixel pitch.

Figure 5.69 shows a cross-sectional view of an actual pixel with 1.0 μm pixel pitch to make it clear how the aspect ratio of the PD is high with a color filter for a single-sensor color system. It is obvious that the PD is quite long vertically. As discussed in Section 2.2, the depth of the PD must be around 3–3.5 μm to absorb enough red light independently from the pixel pitch. Since the thickness of current dye or pigment absorption-type color filters is around 1 μm, the length from the top of the OCL to the PD is never short from the viewpoint of getting incident light to the PD effectively.

Since the shorter focal length of imaging optical elements is employed to achieve thin cameras, the oblique component of the incident ray has been increasing, and rays that pass through color filters that cannot control the direction of propagation enter the adjacent PD, that is, an appreciable extent of cross talk occurs.

In the case of monochrome cameras, cross talk degrades the image sharpness or modulation transfer function of spatial resolution, which can be corrected by signal processing without severe side effects only by edge signal enhancement. On the other hand, the situation caused by cross talk is quite different and more severe in SCCCs. Cross talk causes mixing of not only spatial but also color information in SCCCs. This means inaccurate color reproduction and unnatural color images, which human eyes are sensitive to. Although a certain level of color mixture can be corrected by matrix operations in signal processing, off-diagonal terms in the matrix correspond to mixing levels and are used in calculations for corrections. Since each off-diagonal term already includes noise, noise is reflected to calculated results in amplified or reduced form. Thus, color cross talk often causes substantial noise amplification by the correction operation, and it should be suppressed as much as possible. As described above, the situation in SCCCs is often quite different and much more severe than that in monochrome cameras.

To suppress undesirable cross talk for SCCCs, a light shield film is formed, as shown in Figure 5.69. It is regrettable that BSIs have to give up their 100% aperture ratio for the above reason, despite of so much effort and cost.

Since the cross talk level depends on the degree of oblique incident ray, high-level conditions are required for imaging optical elements such as taking lenses.

As already noted, BSIs have an advantage for making effective use of incident light. However, because the sensing material is silicon, whose absorption coefficient in the visible light region is low, and the color filter is thick, a light shield film is necessary to suppress cross talk. Since it restricts the aperture ratio, it can be said that the potential of BSIs is not currently maximized. Excellent performances are not always achieved with BSIs, but high-level technique is also needed to achieve mastery.

As will be shown in the next section, the advantages of optical performance that appealed at first, such as sensitivity and incident angle dependence, are not maintained compared to advanced FSIs. On the contrary, BSIs have a disadvantage from the viewpoint of cost–benefit performance. This is an example of a technique that has an advantage for monochrome cameras but does not necessarily have the same advantage also for SCCC systems. The differences in sensors for monochrome systems and single-chip color systems should be recognized.

As a recent innovation in BSIs, stacked BSIs[58] have attracted much attention. In conventional BSIs, the supporting substrate plays the role of sustainer, without any other function, although it is made of silicon wafer, as shown in Figure 5.70a. In stacked BSIs, it is composed of a top part (die) and a bottom part (die), as shown in Figure 5.70b. While the pixel array, row drivers, load transistors, and comparators are formed in the top part, the control circuits, row decoder, reference voltage, counter, image processing, and output interface are arranged in the bottom part. They are stacked with the top part up and connected by a through-silicon via (TSV), which is a type of vertical via-type contact. The top part is a CMOS image sensor (CIS) process made of one polysilicon and four metal layers (1P4M), including a color filter and OCL. The bottom part is implemented by a 65 nm 1P7M logic process. In conventional BSIs, an expensive CIS process has to be used for the entire die, but in stacked BSIs, the most appropriate process can be applied to fabricate each die. This is expected to compensate somewhat for the above-mentioned disadvantage of conventional BSIs because of the cost-effective inclusiveness of chip size shrinkage made possible by the CIS process. It is expected that the introduction of various functions will extend this technology to three-dimensionally integrated image sensors. Pixel-level processed image sensors including ADCs might become reality sooner rather than later.

The most advanced BSI and the most advanced FSI are comparable in optical performance such as sensitivity and incident angle dependence, as will be discussed in Section 5.3.3.3.3.

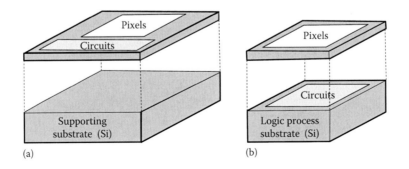

(a) (b)

FIGURE 5.70
Structure of stacked CMOS image sensor: (a) conventional BSI; (b) stacked BSI. (Reprinted with permission from Sukegawa, S., Umebayashi, T., Nakajima, T., Kawanobe, H., Koseki, K., Hirota, I., Haruta, T. et al., *Proceedings of the IEEE International Solid-State Circuits Conference, Digest of Technical Papers*, 27.4, pp. 484–486, 2013.)

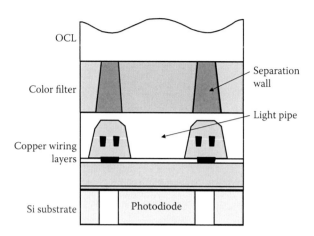

FIGURE 5.71

Cross-sectional view of advanced FSI. (Reprinted with permission from Watanabe, H., Hirai, J., Katsuno, M., Tachikawa, K., Tsuji, S., Kataoka, M., Kawagishi, S. et al., *Proceedings of the IEEE International Electron Devices Meeting, Technical Digest*, 8.3, pp. 179–182, Washington, DC, 2011.)

Stack technology might compensate for the disadvantageous cost-effectiveness of BSI and may achieve high-speed processing.

5.3.3.3.3 *Front Side Illuminated Sensor*

In a recent advanced front side illuminated sensor (SmartFSI®*), it was shown[59] that FSIs achieve performance comparable to that of BSIs. A cross-sectional view with an actual aspect is shown in Figure 5.71.

The pixel pitch is 1.4 μm. The essential point is how effectively incident rays pass through wiring layers and propagate toward the PD and are captured. To this end, the following techniques are adopted: (1) Lightpipe capability is added to the color filter by forming a low refractive index separation wall between each pixel color filter. Therefore, incident light propagates downward. (2) The aperture area is enlarged and the optical stack height is lowered. This is achieved by shrinking the thickness and width of low-resistance copper layers. (3) The original lightpipe is made shallow and powered up due to the high refractive index SiN used on the lightpipe core. (4) A deep PD is formed to effectively capture longer-wavelength light guided to the PD. By using these techniques, quantum efficiency (QE) by simulation of green light (wavelength 520 nm) is 73.9%, while those of conventional FSIs with a deep lightpipe and BSIs with a light shield film are 42.9% and 69.3%, respectively.

Figure 5.72 shows the light energy attenuation along the light propagation direction from the top of the OCL on the extreme right to the silicon surface on the far left. The coordinate value of the abscissa axis shows the height from the silicon surface. The arrows indicate the energy loss through a green filter. The reason they do not start from 100% is that light reflected at the silicon surface and returned to the OCL is subtracted just above the OCL in the simulation. The energy loss through the color filter for the SmartFSI is 3%, while that for the BSI is 14%. The light confinement effect of the separation walls to the color filter prevents energy attenuation at the color filter boundary. The light waves are effectively guided and confined through the color filter between the separation walls and through the lightpipe between the metal apertures and focused on the PD. Peak quantum efficiencies

---

* SmartFSI is a trademark of Panasonic Corporation.

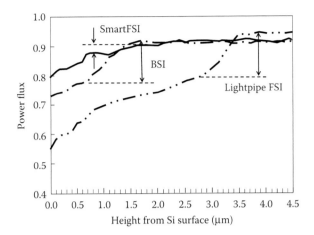

FIGURE 5.72
Simulated light energy loss in light propagating path for green light. Arrows indicate energy loss at color filter. (Reprinted with permission from Watanabe, H., Hirai, J., Katsuno, M., Tachikawa, K., Tsuji, S., Kataoka, M., Kawagishi, S. et al., *Proceedings of the IEEE International Electron Devices Meeting, Technical Digest,* 8.3, pp. 179–182, Washington, DC, 2011.)

of 46.3%, 72.0%, and 63.1% are obtained in the red, green, and blue channels, respectively, as well as low optical and electrical cross talk. The stacked lightpipe structures bring the benefit of higher acceptance angles for incident light, and the maximum incidence angle was 40 ($\pm$20)° at 20% down level of signal quantity of normal light incidence. The situation is the same for smaller pixel pitch of 1.12 µm.[60]

While techniques (1) and (3) seem applicable to BSIs, because they are countermeasures for common issues for both FSIs and BSIs and cost-effectiveness is very important in this industry. Future developments will attract much attention.

5.3.3.4 Organic Sensors

Although organic sensors should not necessarily be categorized as CMOS sensors, they are tentatively classified this way here.

There is a new trend using organic photoconductive material instead of silicon as the sensor part.

Before describing organic sensors, silicon image sensors are reviewed. In Section 2.2, it is pointed out that silicon is not necessarily the best material for the photoelectric conversion part of integrated circuits, although it is a natural choice because it senses visual light. As silicon needs to be 3–4 µm thick to absorb sufficient red light, PDs that are tall or have a high aspect ratio are required along with pixel pitch shrinkage. This structure is apt to bring about cross talk, which should be avoided as much as possible in SCCCs, which is the largest application.

In Section 5.3.2.2, 3-Tr and 4-Tr pixel configurations are compared, and it is pointed out that the CMOS sensor most comparable to CCDs is the 4-Tr type with a combination of a buried/pinned PD. To be pinned, PDs have to be depleted just after signal charges are read out. This means the maximum signal charge, electron, number that can be integrated in a pinned PD is the same as the maximum impurity atom, donor, number in the *n*-type region of the PD. This leads to limitation of the saturation level or dynamic range and inadequate SNR for high-illumination scenes along with pixel pitch shrinkage.

An organic image sensor is one way of overcoming the above limits of silicon image sensors. There are two challenges for 3-Tr pixel configuration. The first is to make extensive changes to improve sensitivity and dynamic range using organic materials as the photoconversion part. The development of the material itself and the fabrication process in combination with the silicon process are included. The other challenge is to overcome the disadvantage of 3-Tr pixel configuration by combining it with organic photoconductive material and creating a new pixel configuration with a noise reduction circuit by using feedback reset.[61]

Organic photoconductive materials have been investigated,[62] but it is indispensable to pursue low-noise-readout techniques to put into practical use.

Some organic materials have an absorption coefficient[63] one order of magnitude higher than that of silicon in the visible region, as shown in Figure 5.73. Therefore, the thickness of photoelectric conversion film (organic photoconductive film [OPF]) can be decreased to <0.5 μm, while in ordinary PDs silicon image sensors are 3–4 μm thick, as shown in Figure 5.74. This makes it possible to remove the light-shielding layer, giving rise to a wide pixel aperture and a wide incident light angle, as shown in Figure 5.74.

Figure 5.75 is a schematic cross-sectional view of a typical pixel region.[64] It consists of micro lenses (OCLs), on-chip color filters (OCF), protective film, the top transparent electrode, OPF, and the bottom pixel electrode directly connected to CMOS circuits. Because of the absence of a light-shielding layer, the ideal aperture ratio of 100% is realized. By applying a positive voltage (V_{top}) to the top transparent electrode, holes of photoconverted charges are collected by the bottom pixel electrode connected to the pixel circuit.

The genuine characteristics of the OPF, QE, and dark current are shown in Figure 5.76. QE measured at a wavelength of 525 nm increases in proportion to V_{top} up to 10 V, beyond which QE saturates. Since the dark current behaves as a monotonously increasing function of V_{top}, the highest SNR is obtained at 10 V.

As mentioned above, an OPF CMOS sensor has 3-Tr pixel configuration and without any countermeasure, it has a large kTC noise of 38 electrons.[62] To solve this problem, a column

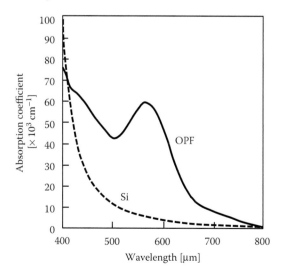

FIGURE 5.73
Absorption coefficient of silicon and OPF. (Reprinted with permission from Isono, S., Satake, T., Hyakushima, T., Taki, K., Sakaida, R., Kishimura, S., Hirao, S. et al., *Proceedings of the 2013 IEEE International Interconnect Technology Conference, paper ID* 3030, 2013.) Copyright [2013] by the Japan Society of Applied Physics.

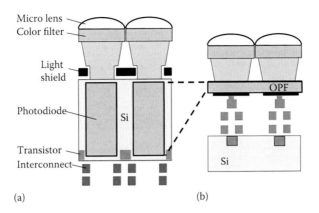

FIGURE 5.74
Comparison of schematic structures: (a) silicon-based PD; (b) OPF-based PD. (Reprinted with permission from Isono, S., Satake, T., Hyakushima, T., Taki, K., Sakaida, R., Kishimura, S., Hirao, S. et al., *Proceedings of the 2013 IEEE International Interconnect Technology Conference, paper ID* 3030, 2013.) Copyright [2013] by the Japan Society of Applied Physics.

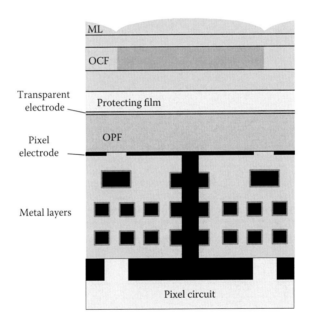

FIGURE 5.75
A cross-sectional schematic of pixels. (Reprinted with permission from Mori, M., Hirose, Y., Segawa, M., Miyanaga, I., Miyagawa, R., Ueda, T., Nara, H. et al., *2013 Symposium on VLSI Technology*, 2–4, pp. 22–24, Kyoto, 2013.) Copyright [2013] by the Japan Society of Applied Physics.

feedback amplifier (FBA) circuit[65] is employed, as shown in Figure 5.77, with a pixel circuit and timing diagram shown in Figure 5.77a and (b), respectively.

An FBA is composed of a differential amplifier and a source follower buffer (not shown in the figure). During the reset period, a tapered reset signal is applied to the reset gate, which lowers the cutoff frequency of the low-pass filter (LPF). A reset operation is carried out by monitoring the signal level, V_{sig}, which is the output of the reset

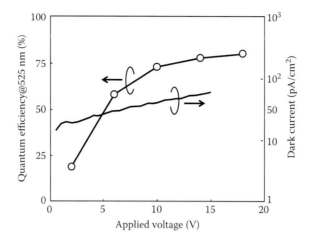

FIGURE 5.76
Quantum efficiency and dark current of OPF as a function of applied voltage. (Reprinted with permission from Mori, M., Hirose, Y., Segawa, M., Miyanaga, I., Miyagawa, R., Ueda, T., Nara, H. et al., *2013 Symposium on VLSI Technology*, 2–4, pp. 22–24, Kyoto, 2013.) Copyright [2013] by the Japan Society of Applied Physics.

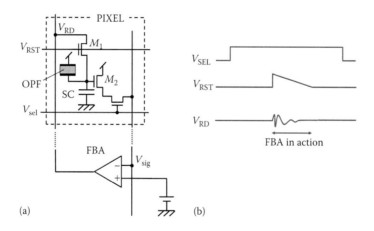

FIGURE 5.77
Noise suppression: (a) schematic diagram of noise-suppressing circuit; (b) timing diagram and operational sequence of noise-suppressing operation. (Reprinted with permission from Ishii, M., Kasuga, S., Yazawa, K., Sakata, Y., Okino, T., Sato, Y., Hirase, J. et al., *2013 Symposium on VLSI Circuits*, 2-3, pp. 8–9, Kyoto, 2013.) Copyright [2013] by the Japan Society of Applied Physics.

level and feeding this back to the reset voltage from FBA output so that the reset level converges to the setup voltage with the timing and operational sequence illustrated in Figure 5.77b. A noise floor <3 electrons is achieved.

In order to extend the saturation level to the highest level possible, a large storage capacitance (SC) is newly incorporated, as shown in Figure 5.77. The storage capacitance allows charge accumulation up to the breakdown voltage of the junction, which is realized by optimizing the impurity profile. Thus, the input of driver transistor M_2 can be swung to the level of the input voltage of M_2 (V_{dd}), giving rise to a much higher saturation level.

Figure 5.78 is a block diagram of an OPF CMOS image sensor.

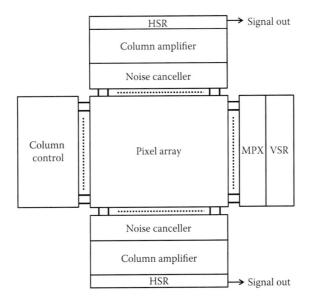

FIGURE 5.78
Block diagram of OPF CMOS image sensor. (Reprinted with permission from Mori, M., Hirose, Y., Segawa, M., Miyanaga, I., Miyagawa, R., Ueda, T., Nara, H. et al., *2013 Symposium on VLSI Technology*, 2–4, pp. 22–24, Kyoto, 2013.) Copyright [2013] by the Japan Society of Applied Physics.

The overall chip performances of three kinds of pixel pitch, 3.0, 1.75, and 0.9 μm, are shown in Table 5.2. For a 3.0 μm pixel, the saturation charge is 77,000 electrons with a V_{dd} of 5.0 V. The dynamic range of the smallest pixel size of 0.9 μm pitch is 68 dB, with a saturation charge of 6,500 electrons. The incident angle with sensitivity reduction of 20% is over 60 (±30)°.

5.4 Electronic Shutter

The basis of the electronic shutter was briefly described in Section 4.3. While it is an important technique to control the amount and length of exposure, there is a big difference between charge transfer–type CCD sensors and XY-address–type MOS and CMOS sensors. Although the exposure timing of all pixels are the same in CCDs, it shifts in series at each row in MOS and CMOS sensors, while exposure time length is the same, as explained by Figure 5.34.

5.4.1 Electronic Shutter of CCD Sensors

In electronic shutter operation in IT-CCDs with a VOD structure, all signal charges integrated in all PDs are discharged to the *n*-type silicon substrate at the start, as shown in Figure 5.23. This operation is called global reset. After discharge completion, signal charge integration, that is, the exposure period, starts at the same time for all pixels. Then, at the end of exposure period, signal charges integrated at all pixels are read out to VCCDs at the same time. Thus, in electronic shutter operation, start timing of exposure is delayed

TABLE 5.2

Chip Performances of Developed Sensors with Different Pixel Sizes

Pixel Size	3.0 µm	1.75 µm	0.9 µm
V_{dd} (V)	5.0		2.8
Number of pixels	999(H)×630(V)	1690(H)×1090(V)	1880(H)×1600(V)
Saturation charge (e⁻)	77,000	60,000	6,500
Sensitivity (e⁻/lux/s)	35,500	12,500	3,200
Readout noise (e⁻)	2.9	2.7	2.5
Dynamic range (dB)	88	87	68
Image lag		Below detection level	

Source: Reprinted with permission from Mori, M., Hirose, Y., Segawa, M., Miyanaga, I., Miyagawa, R., Ueda, T., Nara, H. et al., *2013 Symposium on VLSI Technology*, 2-4, pp. 22–24. Kyoto, 2013. Copyright [2013] by the Japan Society of Applied Physics.

to control the exposure period length by discharging all charges in each PD, and readout timing is kept the same as that of normal exposure mode. The shutter mode with simultaneous exposure timings of all pixels is called global shutter.

5.4.2 Electronic Shutter of MOS and CMOS Sensors

Since reset operations at each PD are done in readout operation in normal exposure mode without an electronic shutter in MOS and CMOS sensors, exposure periods of each row repeat with no interruption, as shown in Figure 5.79a.

As MOS and CMOS sensors are XY-address type, the row select pulse generated by the vertical access circuit such as shift register or decoder is applied to the row select transistor to make it on-state. The readout timing is ruled by the vertical access circuit. In electronic shutter mode, reset operation must be done during the exposure period before the readout timing to set the exposure period, as shown Figure 5.79b. Therefore, a different clock pulse is needed for reset operation from that needed for the sequence of readout operation. Therefore, two series of row accessing signals are necessary. Many CMOS sensors have two vertical access circuits: one is for readout and another is for reset, as shown in Figure 5.80.

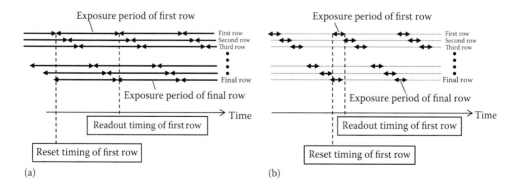

FIGURE 5.79
Exposure period of each row in MOS and CMOS sensors: (a) normal exposure mode; (b) electronic shutter mode.

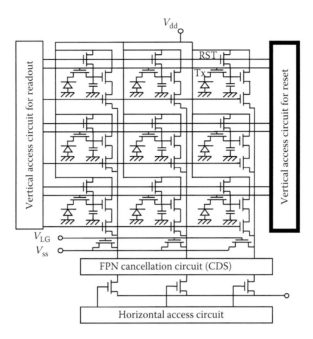

FIGURE 5.80
Schematic diagram of CMOS image sensor with vertical access circuit for electronic shutter.

Since the exposure timing of each row shifts in series, as shown in Figure 5.79b, captured images of objects moving at high speed appear quite different from the real shape because of exposure zone shift with time.

An example of a captured still image, of a rotary fan, is shown in Figure 5.81. Pictures of the still fan and rotating fan captured by a CMOS image sensor in electronic shutter mode are shown in Figure 5.81a and b, respectively. In CMOS sensors, exposure timing shifts from the first row to the final row. Except for the speed, this behavior is the same with focal plane shutters used in single lens reflex cameras. As shown in Figure 5.81c, the exposure zone runs from the top to the bottom. This fan turns in a clockwise direction. It is hoped that readers can visualize how the exposure zone shifts with time on the rotating fan.

In the right half of the fan, the exposure zone and fan blades move in the same direction: downward. This shows that five blades got ahead of the exposure zone during the exposure period. On the other hand, in the left half of the fan, as the exposure zone and fan blades move in opposite directions, many more blades can be seen passing through the exposure zone. Blades seen in both the right and left sides moved from one side to the other during exposure.

It is clear that a shorter time is necessary for the exposure zone to run from the top to the bottom to realize less distortion. This is one of the reasons for the focus on the development of high-speed readout CMOS sensors. The cameras that have a certain level of performance with CMOS sensors are equipped with mechanical shutter systems to avoid these phenomena by controlling the exposure period of whole images simultaneously.

In the case of CCDs, these phenomena do not occur because exposure timings of all pixels are the same, however it causes blur. The electronic shutter mode of CCDs is called global shutter mode while that of CMOS sensors is called rolling shutter mode.

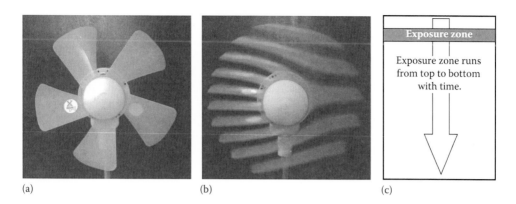

(a) (b) (c)

FIGURE 5.81

Example of still picture of a rotary fan using electronic shutter mode: (a) still object image; (b) still image of rotating fan captured by CMOS sensor; (c) shift of exposure zone with time.

Why can't MOS and CMOS sensors take still images using global shutter mode? In CCDs, signal charges integrated in all pixels are transferred to the VCCD at the same time. This means the VCCD can accept and store signal charges of the same number with pixels; that is, it plays the role of frame memory. On the other hand, in the case of MOS sensors, when signal charges are read out to vertical signal lines, each signal line can accept only one signal charge packet independently at a time, because it is a metallic wiring. Thus, signal lines can accept signal charge packets corresponding to only one row; it has a line memory but not a frame memory as a whole sensor.

How about 4-Tr pixel configuration CMOS sensors? As they have a FD at each pixel, they can play the role of frame memory by transferring a signal charge packet from each PD to each FD in each pixel. But the FD is connected to metallic wiring by ohmic contact with a high enough impurity concentration, and this fact results in higher dark current. Image quality deterioration due to dark current generated during storage in FD is so severe that images obtained this way cannot be practically used. Therefore, in spite of functioning as a frame memory, the performance level is useless. On the other hand, approaches have been proposed that use a similar structure[66] with a buried PD, MOS capacitor,[67] and stacked PIP capacitor for high capacitance in a narrower area with low impurity concentration diffusion[68] for low leakage current, in addition to a MOS capacitor forming out of the image area.

There is another problem for sensors in which the memory part is formed in each pixel. This is the issue of noise due to mixture of charge or light to memory caused by high illumination to neighboring pixels during the signal charge storage period. The mechanism is similar to smear in CCD and MOS sensors. Since image quality debasement due to this phenomenon is severe, efforts to decrease it to a practical use level are being made.

As an aside, the reason smear is not seen in CMOS sensors, unlike CCD and MOS sensors, is not that charges or light do not mix into pixel drains, but because pixel output voltage is decided by the SFA independent of mixture level, that is, unaffected by charge mixture.

5.5 Comparison of Situation and Prospects of Each Sensor Type

Figure 5.82 compares the three types of image sensors from the viewpoint of noise and SNR.

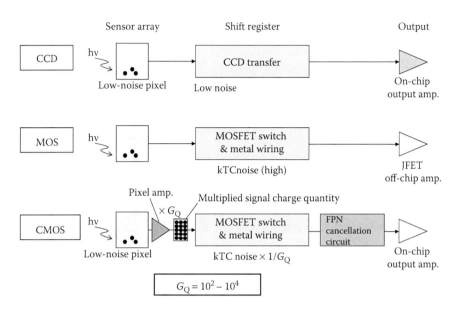

FIGURE 5.82
Comparison from the viewpoint of noise and SNR.

Because of the development of various types of low-noise technologies, in addition to the low-noise performance bestowed on only CCD by complete charge transfer, CCD was the market leader for about two decades from the mid-1980s.

MOS sensors made a grand appearance as a leading competitor in the early 1980s by utilizing LSI processing technology and had a virtual monopoly after their appearance on the market.

They suffered from high-level kTC noise due to the vertical signal line. While MOS sensors banished kTC noise by removing the vertical signal line itself in the TSL sensor, improvement of SNR was restricted by the temporal noise of the off-chip amplifier because of the current readout. As a result, MOS sensors could not compete with CCDs when they were launched on the market later.

CMOS sensors did not have the opportunity to appear on the market until the age of progress in fine-pitch transistor technology, such as 0.35 μm transistor process technology enough to form three or four transistors in a pixel, in the late 1990s. In the mid-2000s, widely used cell phone cameras equipped with CMOS sensors were given a low evaluation. The picture quality of CMOS sensors was quite low compared to that of CCDs. But CMOS sensors originally aspired to suppress noise impact by charge quantity multiplication at pixel level before accepting noise at the next stage, and to cancel FPN caused by variation of pixel amplifier characteristics. Therefore, CMOS sensors became comparable to CCDs as a result of the realization of low-noise PDs by importing pixel technologies amassed through CCDs and achievement of suppression of noise impact to a negligible level by large increases in the charge quantity of the pixel amplifier. Sensors with a noise electron number less than unity were reported. This was achieved by additional corrections of FPN due to variations of column circuit characteristics and procedures of circuit technology such as column amplifiers and column A/D converters. Thus, CMOS sensors have a SNR advantage over CCDs. It is

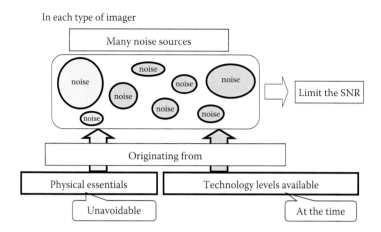

FIGURE 5.83
What should be taken into account when choosing which technologies to adopt.

hoped further progress will be made using stacked BSIs, advanced FSIs, and organic sensors.

In CCDs, noise reduction pixel techniques were pursued to make good use of low-noise performance due to complete charge transfer. It was a technological necessity that CMOS sensors had low-noise pixels in the same way as CCDs, because they aim for high SNR through high multiplication gain of signal charge quantity by pixel amplifier.

While CCDs were a major player in the imaging field for about two decades, CMOS sensors have made advances in low-noise performance. The present situation of CCDs is that they can capture images with quality suitable for practical use at reasonable cost and realize global shutter mode at no additional cost.

CCDs are of good birth and beauty without makeup: they use only basic cosmetology such as device structures and process technologies. On the other hand, CMOS sensors have progressed by using basic cosmetology, that is, the pixel technology of CCDs. This means that high-performance CMOS sensors cannot be manufactured by a low-cost vanilla CMOS logic process. It could be said that the beauty of CMOS sensors is further evolving with use of cosmetic surgery, namely circuit technology.

Figure 5.83 summarizes what should be taken into account when technologies are chosen and is not restricted to image sensors. Each type of image sensor has many noise origins that restrict the SNR. Some of them originate from unavoidable physical essentials, such as kTC noise, and others come from the level of technology available at the time. The most important thing to evaluate is which ones are essential, because noise is the main factor that determines what level of performance can be reached, which dictates the value of the image sensor.

In MOS and CMOS sensors, MOS and CMOS technology are available. MOS sensors have become a leading competitor because of this. Both sensors can share in the bounty of the progress of MOS/CMOS technologies, such as transistor size shrinkage, low power, and low voltage. As the column-parallel ADC in CMOS sensors shows, a lot is expected from on-chip functional circuits. An example is the one-chip camera[69] named ASIC VISION,[70] realized with a pioneering spirit in 1990.

References

1. T. Kuroda, Japan in the world of image sensors—What Japan has been playing the role of, and what could be learnt from the history, *Journal of ITE*, 65(3), 336–341, 2011. https://www.jstage.jst.go.jp/article/itej/65/3/65_3_336/_pdf (accessed January 10, 2014).
2. W. S. Boyle, G. E. Smith, Charge coupled semiconductor devices, *Bell System Technical Journal*, 49, 587–593, 1970.
3. R. H. Walden, R. H. Krambeck, R. J. Strain, J. McKenma, N. L. Schryer, G. E. Smith, The buried channel charge coupled device, *Bell System Technical Journal*, 51, 1635–1640, 1972.
4. M. F. Tompsett, G. F. Amelio, W. J. Bertran, R. R. Buckley, W. J. Mcnamara, J. C. Mikkelsen, D. A. Sealer, Charge-coupled imaging devices: Experimental results, *Transactions on Electron Device*, 18(11), 992–996, 1971.
5. G. F. Amelio, Physics and applications of charge-coupled devices, *IEEE Intercon*, 1/3, 1973.
6. S. Okamoto, T. Kuroda, Japan patent, 3303384 (May 10, 2002).
7. Y. Sone, K. Ishikawa, S. Hashimoto, T. Kuroda, Y. Ohkubo, A single chip CCD color camera system using field integration mode, *Journal of the Institute of Television Engineers of Japan*, 37(10), 855–862, 1983.
8. Y. Ishihara, K. Tanigaki, A high photosensitivity IL-CCD image sensor with monolithic resin lens array, in *Proceedings of the International Electron Devices Meeting*, 19-3, pp. 497–500, 1983, Washington, DC.
9. A. Tsukamoto, W. Kamisaka, H. Senda, H. Niisoe, H. Aoki, T. Otagaki, Y. Shigeta, et al., High sensitivity pixel technology for a 1/4-inch PAL 430 k pixel IT-CCD, in *Proceedings of the IEEE Custom Integrated Circuits Conference*, pp. 39–42. 5–8 May 1996, San Diego, CA.
10. C. H. Sequin, Blooming suppression in charge coupled area imaging, *Bell System Technical Journal*, 51, 1923–1926, 1972. http://www.alcatel.hu/bstj/vol51-1972/articles/bstj51-8-1923.pdf (accessed January 10, 2014).
11. Y. Ishihara, E. Oda, H. Tanigawa, A. Kohno, N. Teranishi, E. Takeuchi, I. Akiyama, T. Kamata, Interline CCD image sensor with an anti blooming structure, in *Proceedings of the IEEE International Solid-State Circuits Conference Digest of Technical Papers*, pp. 168–169. 10–12 February 1982, San Francisco, CA.
12. T. Kuroda, K. Horii, S. Matsumoto, Y. Hiroshima, High performance interline transfer CCD, *National Technical Report*, 28(2), 12: 107–116, 1982.
13. N. Teranishi, A. Kohno, Y. Ishihara, E. Oda, K. Arai, No image lag photodiode structure in the interline CCD image sensor, *IEDM Digest of Technical Papers*, 12(6), 324–327, 1982.
14. T. Kuroda, K. Horii, Japan patent, 1996064 (March 13, 1992).
15. T. Kozono, T. Kuroda, Y. Matsuda, T. Yamaguchi, T. Kuriyama, S. Matsumoto, E. Fujii, Y. Hiroshima, K. Horii, Small image size CCD sensor, ITE Technical Report, TEBS 109-5, pp. 25–30, 1986.
16. B. C. Burkey, W. C. Chang, J. Littlehale, T. H. Lee, T. J. Tredwell, J. P. Lavine, E. A. Trabka, The pinned photodiode for an interline-transfer CCD image sensor, *IEDM Digest of Technical Papers*, 2(3), 28–31, 1984.
17. K. Horii, T. Kuroda, T. Kunii, A new configuration of CCD imager with a very low smear level, *Electron Device Letters*, EDL-2, 12, 1981.
18. G. P. Weckler, A silicon photodevice to operate in a photon flux integration mode, in *Proceedings of International Electron Devices Meeting Digest of Technical Papers*, 10-5, October 1965, Washington, DC.
19. G. P. Weckler, Operation of p-n junction photodetectors in a photon flux integrating mode, *Journal of Solid-State Circuits*, 2, 65–73, 1967.
20. M. Masuda, M. Noda, Y. Saito, S. Ohba, K. Takahashi, T. Fujita, Solid state color camera with single-chip MOS imager, ITE Technical Report, 4, 41, TEBC69-1, pp. 1–6, February 1981.
21. I. Takemoto, MOS imager, *Journal of ITE*, 40(11), 1067–1072, 1986.

22. M. Aoki, H. Ando, S. Ohba, I. Takemoto, S. Nagahara, T. Nakano, M. Kubo, T. Fujita, 2/3" inch format MOS single-chip color imager, *IEEE Transactions on Electron Devices*, ED-29(4), 745–750, 1982.

23. K. Takahashi, N. Ozawa, M. Aoki, I. Takemoto, T. Suzuki, T. Miyazawa, S. Nagahara, High resolution MOS area imaging device, ITE Technical Report, 9, 17, ED-623, pp. 13–18, February 1982.

24. K. Takahashi, S. Nagahara, I. Takemoto, M. Aoki, N. Ozawa, T. Suzuki, High resolution MOS area sensor, *Journal of the Institute of Television Engineers of Japan*, 37(10), 812–818, 1983.

25. I. Takemoto, T. Miyazawa, S. Nishizawa, M. Uehara, M. Nakai, T. Akiyama, TSL (transversal signal line) solid state imager, ITE Technical Report, 9, 17, ED-891, pp. 49–54, September 1985.

26. P. Noble, Self-scanned silicon image detector arrays, *Transactions on Electron Devices*, ED-15, 202–209, 1968.

27. F. Andoh, J. Matsuzaki, An imaging device with amplifier in each photo sensing element, in *Proceeding of IEICE General Conference*, No.1159, 1981, Tokyo.

28. F. Andoh, K. Taketoshi, J. Yamazaki, M. Sugawara, Y. Fujita, K. Mitani, Y. Matuzawa, K. Miyata, S. Araki, A 250000-pixel image sensor with FET amplification at each pixel for high-speed television cameras, in *Proceedings of the IEEE International Solid-State Circuits Conference Digest of Technical Papers*, pp. 212–213. 14–16 February 1990, San Francisco, CA.

29. A. Theuwissen, 50 years of solid-state image sensors at ISSCC, in *Proceedings of the IEEE International Solid-State Circuits Conference Digest of Technical Papers*, S26, 2003, San Francisco, CA.

30. N. Tanaka, S. Hashimoto, M. Shinohara, S. Sugawa, M. Morishita, S. Matsumoto, Y. Nakamura, T. Ohmi, A 310 k pixel bipolar imager (BASIS), in *Proceedings of the IEEE International Solid-State Circuits Conference Digest of Technical Papers*, WPM 8.5, pp. 96–97, 1989, New York.

31. J. Hynecek, A new device architecture suitable for high-resolution and high-performance image sensors, *Transactions on Electron Devices*, 35(5), 646–652, 1988.

32. E. Oba, K. Mabuchi, Y. Lida, N. Nakamura, H. Miura, A 1/4 inch 330 k square pixel progressive scan CMOS active pixel image sensor, in *Proceedings of the IEEE International Solid-State Circuits Conference Digest of Technical Papers*, pp. 180–181, 1997, San Francisco, CA.

33. Y. Matsunaga, Y. Endo, Noise cancel circuit for CMOS image sensor, *ITE Technical Report*, 22(3), 7–11, 1998.

34. M. Mori, M. Katsuno, S. Kasuga, T. Murata, T. Yamaguchi, A 1/4in 2M pixel CMOS image sensor with 1.75 transistor/pixel, in *Proceedings of the IEEE International Solid-State Circuits Conference Digest of Technical Papers*, 6.2, pp. 80–81, 2004, San Francisco, CA.

35. K. Mabuchi, N. Nakamura, E. Funatsu, T. Abe, T. Umeda, T. Hoshino, R. Suzuki, H. Sumi, CMOS image sensor using a floating diffusion driving buried photodiode, in *Proceedings of the IEEE International Solid-State Circuits Conference Digest of Technical Papers*, 6.3, pp. 82–83, 2004, San Francisco, CA.

36. H. Takahashi, M. Kinoshita, K. Morita, T. Shirai, T. Sato, T. Kimura, H. Yuzurihara, S. Inoue, A 3.9 µm pixel pitch VGA format 10b digital image sensor with 1.5-transistor/pixel, in *Proceedings of the IEEE International Solid-State Circuits Conference Digest of Technical Papers*, 6.1, pp. 108–109, 2004, San Francisco, CA.

37. K. Itonaga, K. Mizuta, T. Kataoka, M. Yanagita, H. Ikeda, H. Ishiwata, Y. Tanaka, et al., Extremely-low-noise CMOS image sensor with high saturation capacity, in *Proceedings of the IEEE International Electron Devices Meeting Digest of Technical Papers*, 8.1.1–8.1.4, pp. 171–174. 5–7 December 2011, Washington, DC.

38. S. Kawahito, M. Sakakibara, D. Handoko, N. Nakamura, H. Satoh, M. Higashi, K. Mabuchi, H. Sumi, A column-based pixel-gain-adaptive CMOS image sensor for low-light-level imaging, in *Proceedings of the IEEE International Solid-State Circuits Conference Digest of Technical Papers*, 12.7, pp. 224–225, 2003, San Francisco, CA.

39. H. Takahashi, T. Noda, T. Matsuda, T. Watanabe, M. Shinohara, T. Endo, S. Takimoto, et al., A 1/2.7 inch low-noise CMOS image sensor for full HD camcorders, in *Proceedings of the IEEE International Solid-State Circuits Conference Digest of Technical Papers*, 28.6, pp. 510–511. 11–15 February 2007, San Francisco, CA.

40. B. Pain, E. Fossum, Approaches and analysis for on-focal-plane analog-to-digital conversion, in *Proceedings of the SPIE, Volume 2226, Aerospace Sensing—Infrared Readout Electronics* II, pp. 1–11, 1994, San Diego, CA.

41. B. Fowler, D. Yang, A. E. Gamal, Techniques for pixel level analog to digital conversion, *Aerosense*, 98, 3360–3361, 1998. http://www-isl.stanford.edu/~abbas/group/papers_and_pub/aerosense98_slide.pdf (accessed January 10, 2014).

42. F. Andoh, M. Nakayaka, H. Shimamoto, Y. Fujita, A digital pixel image sensor with 1-bit ADC and 8-bit pulse counter in each pixel, IISW, P1, (1999). http://www.imagesensors.org/Past%20Workshops/1999%20Workshop/1999%20Papers/12%20Andoh%20et%20al.pdf (accessed January 10, 2014).

43. K. Cho, C. Lee, S. Eikedal, A. Baum, J. Jiang, C. Xul, X. Fan, R. Kauffman, A 1/2.5 inch 8.1 Mpixel CMOS image sensor for digital cameras, in *Proceedings of the IEEE International Solid-State Circuits Conference Digest of Technical Papers*, 28.5, pp. 508–509, 2007, San Francisco, CA.

44. Y. Nitta, Y. Muramatsu, K. Amano, T. Toyama, J. Yamamoto, K. Mishina, A. Suzuki, et al., High-speed digital double sampling with analog CDS on column parallel ADC architecture for low-noise active pixel sensor, in *Proceedings of the IEEE International Solid-State Circuits Conference Digest of Technical Papers*, 27.5, pp. 2024–2031. 6–9 February 2006, San Francisco, CA.

45. S. Yoshihara, M. Kikuchi, Y. Ito, Y. Inada, S. Kuramochi, H. Wakabayashi, M. Okano, et al., A 1/1.8-inch 6.4 M Pixel 60 frames/s CMOS image sensor with seamless mode change, in *Proceedings of the IEEE International Solid-State Circuits Conference Digest of Technical Papers*, 27.1, pp. 1984–1993. 6–9 February 2006, San Francisco, CA.

46. S. Matsuo, T. J. Bales, M. Shoda, S. Osawa, K. Kawamura, A. Andersson, M. Haque, et al., 8.9-megapixel video image sensor with 14-b column-parallel SA-ADC, *Transactions on Electron Devices*, 56(11), 2380–2389, 2009.

47. J. Park, S. Aoyama, T. Watanabe, K. Isobe, S. Kawahito, A high-speed low-noise CMOS image sensor with 13-b column-parallel single-ended cyclic ADCs, *Transactions on Electron Devices*, 56(11), 2414–2422, 2009.

48. J. Park, S. Aoyama, T. Watanabe, T. Akahori, T. Kosugi, K. Isobe, Y. Kaneko, et al., A 0.1e- vertical FPN 4.7e- read noise 71 dB DR CMOS image sensor with 13b column-parallel single-ended cyclic ADCs, in *Proceedings of the IEEE International Solid-State Circuits Conference Digest of Technical Papers*, 15.3, pp. 268–269. 8–12 February 2009, San Francisco, CA.

49. Y. Chae, J. Cheon, S. Lim, D. Lee, M. Kwon, K. Yoo, W. Jung, D. Lee, S. Ham, G. Han, A 2.1 M pixel 120 frame/s CMOS image sensor with column-parallel ΔΣ ADC architecture, in *Proceedings of the IEEE International Solid-State Circuits Conference Digest of Technical Papers*, 22.1, pp. 394–395. 7–11 February 2010, San Francisco, CA.

50. Y. Chae, J. Cheon, S. Lim, M. Kwon, K. Yoo, W. Jung, D. Lee, S. Ham, G. Han, A 2.1 M pixels, 120 frame/s CMOS image sensor with column-parallel 16 ADC architecture, *Journal of Solid State Circuits*, 46(1), 236–247, 2011.

51. W. Yang, A wide-dynamic-range, low-power photosensor array, in *Proceedings of the IEEE International Solid-State Circuits Conference, 41st ISSCC Digest of Technical Papers*, 13.7, pp. 230–231. 16–18 February 1994, San Francisco, CA.

52. J. Gambino, B. Leidy, J. Adkisson, M. Jaffe, R. J. Rassel, J. Wynne, J. Ellis-Monaghan, et al., CMOS imager with copper wiring and lightpipe, in *Proceedings of the International Electron Devices Meeting, Technical Digest*, pp. 5.5.1–5.5.4. 11–13 December 2006, San Francisco, CA.

53. J. Tower, P. Swain, F. Hsueh, R. Dawson, P. Levine, G. Meray, J. Andrews, et al., Large format backside illuminated CCD imager for space surveillance, *Transactions on Electron Devices*, 50(1), 218–224, 2003.

54. S. Iwabuchi, Y. Maruyama, Y. Ohgishi, M. Muramatsu, N. Karasawa, T. Hirayama, A back-illuminated high-sensitivity small-pixel color CMOS image sensor with flexible layout of metal wiring, in *Proceedings of the IEEE International Solid-State Circuits Conference Digest of Technical Papers*, 16.8, pp. 1171–1178. 6–9 February 2006, San Francisco, CA.

55. Papers in IISW 2009, Backside Illumination Symposium, IISW (June 2009). http://www.imag-esensors.org/Past%20Workshops/2009%20Workshop/2009%20Papers/2009%20IISW%20 Program.htm (accessed January 10, 2014).

56. M. Nakai, K. Watanabe, I. Takemoto, S. Shimada, T. Nagano, Si substrate for low noise color image sensor, in *Proceeding of ITE General Conference*, pp. 21–22, 1982, Tokyo.

57. Y. Hiroshima, T. Kuroda, S. Matsumoto, K. Horii, T. Kunii, CCD image sensor fabricated on EPI-wafers, IEICE Technical Report, SSD81-129, pp. 103–108, February 1982.

58. S. Sukegawa, T. Umebayashi, T. Nakajima, H. Kawanobe, K. Koseki, I. Hirota, T. Haruta, et al., A 1/4-inch 8 M pixel back-illuminated stacked CMOS image sensor, in *Proceedings of the IEEE International Solid-State Circuits Conference Digest of Technical Papers*, 27.4, pp. 484–486, 2013, San Francisco, CA.

59. H. Watanabe, J. Hirai, M. Katsuno, K. Tachikawa, S. Tsuji, M. Kataoka, S. Kawagishi, et al., A 1.4 µm front-side illuminated image sensor with novel light guiding structure consisting of stacked lightpipes, in *Proceedings of the IEEE International Electron Devices Meeting, Technical Digest*, 8.3, pp. 179–182. 5–7 December 2011, Washington, DC.

60. N. Teranishi, H. Watanabe, T. Ueda, N. Sengoku, Evolution of optical structure in image sensors, in *Proceedings of the IEEE International Electron Devices Meeting, Technical Digest*, 24.1, pp. 533–536. 10–13 December 2012, San Francisco, CA.

61. T. Kuroda, M. Masuyama, U.S. Patent 6469740.

62. M. Ihama, H. Inomata, H. Asano, S. Imai, T. Mitsui, Y. Imada, M. Hayashi, et al., CMOS image sensor with an overlaid organic photoelectric conversion layer, IISW, P33, 2011.

63. S. Isono, T. Satake, T. Hyakushima, K. Taki, R. Sakaida, S. Kishimura, S. Hirao, et al., A 0.9 µm pixel size image sensor realized by introducing organic photoconductive film into the BEOL process, in *Proceedings of the 2013 IEEE International Interconnect Technology Conference, paper ID* 3030, 2013, Hokkaido, Japan.

64. M. Mori, Y. Hirose, M. Segawa, I. Miyanaga, R. Miyagawa, T. Ueda, H. Nara, et al., Thin organic photoconductive film image sensors with extremely high saturation of 8500 electrons/µm², in *2013 Symposium on VLSI Technology*, 2-4, pp. 22–24. 11–13 June 2013, Kyoto.

65. M. Ishii, S. Kasuga, K. Yazawa, Y. Sakata, T. Okino, Y. Sato, J. Hirase, et al., An ultra-low noise photoconductive film image sensor with a high-speed column feedback amplifier noise cancel-ler, in *2013 Symposium on VLSI Circuits*, 2-3, pp. 8–9. 12–14 June 2013, Kyoto.

66. K. Yasutomi, S. Itoh, S. Kawahito, A 2.7e—Temporal noise 99.7% shutter efficiency 92 dB dynamic range CMOS image sensor with dual global shutter pixels, in *Proceedings of the IEEE International Solid-State Circuits Conference Digest of Technical Papers*, 22.3, pp. 398–399, 2010, San Francisco, CA.

67. M. Sakakibara, Y. Oike, T. Takatsuka, A. Kato, K. Honda, T. Taura, T. Machida, et al., An 83 dB-dynamic-range single-exposure global-shutter CMOS image sensor with in-pixel dual storage, in *Proceedings of the IEEE International Solid-State Circuits Conference Digest of Technical Papers*, 22.1, pp. 380–382. 19–23 February 2012, San Francisco, CA.

68. Y. Tochigi, K. Hanzawa, Y. Kato, R. Kuroda, H. Mutoh, R. Hirose, H. Tominaga, K. Takubo, Y. Kondo, S. Sugawa, A global-shutter CMOS image sensor with readout speed of 1 T pixel/s burst and 780 M pixel/s continuous, in *Proceedings of the IEEE International Solid-State Circuits Conference Digest of Technical Papers*, 22.2, pp. 382–384. 19–23 February 2012.

69. S. G. Smith, J. E. D. Hurwitz, M. J. Torrie, D. J. Baxter, A. A. Murray, P. Likoudis, A. J. Holmes, et al., A single-chip CMOS 306 × 244-pixel NTSC video camera and a descendant coprocessor device, *Journal of Solid State Circuits*, 33(12), 2104–2111, 1998.

70. D. Renshaw, P. B. Denyer, G. Wang, M. Lu, ASIC vision, in *Proceedings of the IEEE 1990 Custom Integrated Circuits Conference*, pp. 7.3.1–7.3.4, May 1990, Boston, MA.

6

Impacts of Digitization by Built-In Coordinate Points on Image Information Quality

Chapter 1 described the factors that compose image information (light intensity, space [position], wavelength, time) and how all factors except light intensity are built-in coordinate points in imaging systems. Image sensors measure the number of photons entering the domain of each built-in coordinate point. This means that sensors integrate the signal charge generated when an incident photon reaches three territories of each coordinate point, that is, at each pixel area, through each color filter, and during the exposure period of each frame. This operation is known as sampling of photon numbers at each coordinate point. This chapter discusses the impacts of the sampling operation at each digitized built-in coordinate point.

6.1 Sampling and Sampling Theorem

A specific example used here is space sampling. Sensors integrate signal charges generated by incident light* that enters the sensor part formed in each pixel arranged in a two-dimensional area. The output signal at each pixel during one exposure period is only one, and the signal value is the incident light intensity information at the coordinate point. Therefore, if the number of pixels is smaller, or the periodicity of sampling or the space frequency are lower, then the image quality based on space information is low because of coarse sampling, as shown in Figure 6.1.

Figure 6.2 explains how the spatial frequency in the obtained image information is restricted. The solid line in the top frame indicates three kinds of frequency and four input signals. A sine wave curve whose frequency is sampling frequency, f_s, is shown in the bottom frame. The sampling pitch p is expressed as $1/f_s$. Sampling operations are carried out at the positions of the maximum point, as indicated by the up arrows. In the case that the input signal frequency is sufficiently low compared with the sampling frequency, f_s, as shown in Figure 6.2d, a broken curve obtained by tracing the sampling point, which is indicated by filled circles, accurately shows the same wave as the input signal. Both the amplitude and the frequency are maintained.

Then how is it possible to reproduce high frequency? The case when the frequency is half the sampling frequency, that is, $f = f_s/2$, is shown in Figure 6.2c. In this figure, the positions of the peaks and troughs (corresponding to white and black in the images) fit with that of the sampling points shown by filled circles. The amplitude and frequency of the reproduced curve shown by a broken line are retained, although the shape is a triangular waveform. Because this is the condition in which peaks and troughs fit with the sampling

---

* Actually, this is light that passes through the color filter of a pixel.

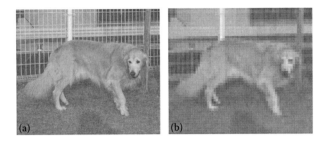

FIGURE 6.1   (See color insert)
Comparison of sampling frequency dependence of spatial information quality: (a) fine sampling with 1318×1106 pixels; (b) course sampling with 64×54 pixels (without smoothing).

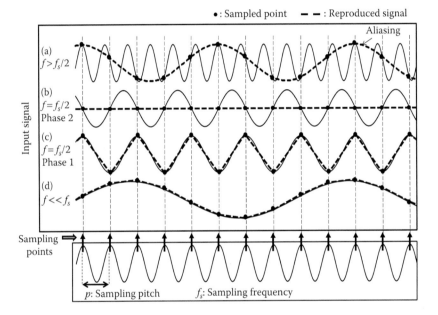

FIGURE 6.2
Sampling and sampling theorem.

points, it is easily understood that any input signal with a higher frequency than $f_s/2$ cannot be reproduced accurately by the sampling frequency f_s. Therefore, the maximum frequency that can be reproduced is just half the sampling frequency. This frequency is called the Nyquist frequency, and the relation is called the Nyquist theorem or the sampling theorem.

Denoting the sampling pitch and the Nyquist frequency as p and f_N, respectively, we obtain the following relations:

$$f_N = \frac{1}{2}f_s = \frac{1}{2p} \tag{6.1}$$

However, an input signal whose frequency is $f_s/2$ is not always reproduced accurately. Figure 6.2b shows an input signal whose frequency is $f_s/2$, which is the same as Figure 6.2c, the only difference being that the phase is a quarter cycle late. The positions of the sampling

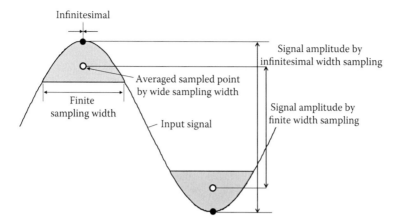

FIGURE 6.3
Sampling width and the sampled signal amplitude.

points are in the middle of the peaks and troughs (corresponding to gray in the images). The signal curve that is obtained by tracing the sampled points is flat with no amplitude, as shown in Figure 6.2b. In this case, neither the amplitude nor the frequency is retained. Thus, this phase is also important, especially around the Nyquist frequency, as will be seen.

An input signal whose frequency is higher than the Nyquist frequency is shown in Figure 6.2a. The reproduced signal obtained by tracing the sampled points is very different from the original input signal, as shown in the figure. This false signal is called aliasing or folding noise. Thus, a higher sampling frequency is necessary to obtain accurate signal information for a higher Nyquist frequency.

In the above description, the sampling width is assumed to be infinitesimal. However, the actual sampling operation cannot be realized with a zero sampling width, but requires some finite width. Figure 6.3 shows the impact of the sampling width on the sampled result.

In the case of an infinitesimal sampling width, the maximum and minimum values of the input signal are reflected in the sampled result. However, in the case of sampling with a finite width, the sampling is carried out during the sampling period by integration or averaging. Thus, the maximum and minimum values cannot be directly reflected as sampled points. Accordingly, a larger amplitude of the sampled signal is obtained by sampling with a narrower sampling width.

6.2 Sampling in Space Domain

A schematic diagram of spatial sampling is shown in Figure 6.4a. Pixels are periodically arrayed with sampling pitch p and aperture a in real space. Since only light that passes through an aperture can reach the sensor parts, the sampling operation is only carried out in the aperture area; that is, the aperture width is the same as the sampling width. As the sampling pitch is p, the sampling frequency f_s equals $1/p$ and the sampling width is a. Under this condition, the frequency dependency of the sampled signal amplitude of sine wave input signals is shown in Figure 6.4b. The frequency is normalized by the sampling

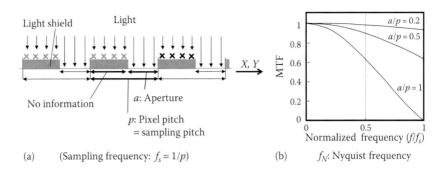

(a)   (Sampling frequency: $f_s = 1/p$)   (b) f_N: Nyquist frequency

FIGURE 6.4
Space sampling: (a) schematic diagram of space sampling by pixel aperture; (b) frequency dependence of MTF.

frequency. The three states of aperture pitch, a/p, which are 1, 0.5, and 0.2, normalized by the sampling pitch p are shown. As mentioned in Section 6.1, a wider aperture indicates a lower amplitude, especially in the higher-frequency region. This signal amplitude shows the spatial frequency response characteristics in transfer systems and is called the modulation transfer function (MTF). To obtain higher-frequency information, a higher Nyquist frequency, that is, a shorter pixel pitch, is required, as shown in Figure 6.1. Although a narrower aperture gives a higher amplitude, amplitude performance is not emphasized in usual applications because the narrower aperture brings about lower sensitivity, while higher sensitivity is the first preference for imaging systems.

Next, the images that are obtained by spatial sampling are confirmed by simple simulations using spreadsheet software. Using the inputs, the signal is observed using a circular zone plate (CZP) chart, which is often used to check the frequency of false signals. A calculated drawing in mathematical form is shown in Figure 6.5, indicating in a concentric

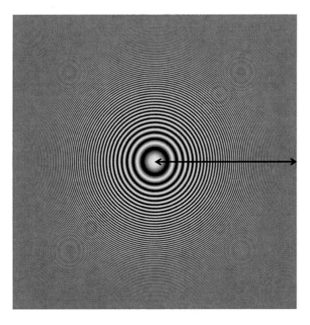

FIGURE 6.5
CZP chart (calculated drawing).

fashion that the spatial frequency is in proportion with the square of the distance from the center. The waveform from the center to the right edge along the arrow indicated in Figure 6.5 is shown in Figure 6.6a.

Figure 6.6b shows aperture periodicity, that is, the sampling frequency. To emphasize the effects, a coarser pitch is chosen, and the aperture ratio a/p is set at 0.5. As the waveform in Figure 6.6a is sampled with the pitch of Figure 6.6b, the mathematical forms of Figure 6.6a and b are multiplied using spreadsheet software. All values except those for the aperture areas in Figure 6.6b are set to zero. The calculated results are shown in Figure 6.6c and d. That the amplitude modulation is seen even in a lower-frequency region than the Nyquist frequency indicates that the amplitude is not reproduced accurately according to the sampling conditions. Figure 6.6c and d in a higher area than the Nyquist frequency show that the waveforms are false signals completely different from the input signals. Specifically, Figure 6.6c shows symmetry with the Nyquist frequency at the axis, as the name "folding noise" suggests.

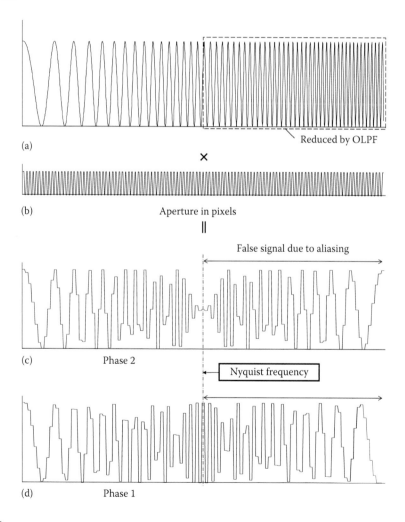

FIGURE 6.6
Model simulation of CZP pattern using spreadsheet software: (a) waveform of CZP pattern input signal; (b) sampling pitch; (c, d) calculated results.

In actual imaging systems, the signal component at and over the Nyquist frequency is removed or reduced by using an optical low-pass filter (OLPF) to avoid impacts caused by false signals, as indicated in Figure 6.6a. The difference between Figure 6.6c and d is the sampling phase. While the amplitude at the Nyquist frequency is retained in Figure 6.6c, it is zero in Figure 6.6d. This result is due to the difference between the sampling phases, as explained in Figure 6.2. The sampling phases in Figure 6.6c and d correspond to those of Figure 6.2b and c, respectively.

Figure 6.7a and b show real pictures of a CZP chart taken by image sensors. In the CZP chart, the resolutions at the right and left edges of the line passing through the center mean 600 television (TV) lines,* while the top and bottom edges correspond to 450 TV lines.

Figure 6.7a shows an emphasized picture of a CZP chart taken with a CCD with a 4.1 μm pitch square pixel with 955(H) × 550(V) numbers without an OLPF. Figure 6.7a shows many false signals of concentric circles, especially the strong signal observed at the Nyquist frequency of 550 TV lines in both the picture and the measured signal amplitude. Figure 6.7b is a picture taken with the OLPF set just in front of the sensor to suppress the false signal. The false signal is suppressed to an unobservable level in the picture and the amplitude shows almost zero at the Nyquist frequency. Comparing the amplitude graphs of Figure 6.7a and b, it can be seen in Figure 6.7b that the false signal at the Nyquist frequency is removed completely and the amplitude decreases with the frequency, especially at areas higher than the Nyquist frequency by the effect of OLPF.

An OLPF is a low-pass filter of spatial frequency, as its name indicates.[1,2] The most commonly used base material for OLPFs is crystalline quartz. Using birefringence of the crystal, the incident light beam is split into two parts, an ordinary ray and an extraordinary ray, as shown in Figure 6.8. While the ordinary ray propagates to the pixel directly underneath, the extraordinary ray is one-pixel pitch shifted[†] through the crystal and, accordingly, reaches the pixel next to the pixel that the ordinary beam arrives at, as shown.

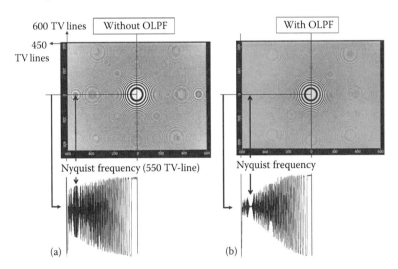

FIGURE 6.7
Examples of pictures of a CZP chart taken at 4.1 μm pitch with a 955(H) × 550(V) pixel CCD: (a) without OLPF; (b) with OLPF.

---

* In a TV line expression system, a pair of black and white lines is counted as two lines.
† The thickness of the OLPF is chosen so that the shift distance through the crystal equals one-pixel pitch.

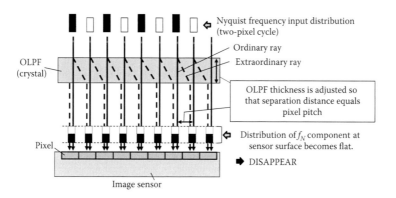

FIGURE 6.8
Principle operation of an optical low-pass filter.

Because one cycle of the Nyquist frequency input is a two-pixel pitch, it has the spatial frequency shown by the black and white bars in Figure 6.8. Each intensity of the bars is divided in half: one half is underneath the pixel and the other adjoins the pixel. (The OLPF thickness is adjusted so that the separation distance equals the pixel pitch.) Therefore, each pixel receives a light intensity of half a black bar and half a white bar of the Nyquist frequency component of input, shown at the top of the figure. This means that the intensity of this frequency component is distributed equally to each pixel, or no amplitude. As the component of the Nyquist frequency vanishes through the OLPF in this way, the false signal is greatly reduced, as shown in Figure 6.7b.

As can be understood from the mechanism, the frequency whose amplitude is deleted perfectly is only one point, the effect remains around the target frequency to reduce the signal amplitude, as shown in the bottom graph in Figure 6.7b. Thus, the OLPF deletes the Nyquist frequency component, which causes a false signal and reduces the amplitude around it and the higher-frequency component of it.

The case of a smaller pixel pitch is discussed. Figure 6.9 shows pictures of a CZP chart taken by an image sensor at 1.8 µm pitch and 3096(H) × 2328(V) pixels without an OLPF. Since the resolution of the CZP with a full angle of view is only 600 TV lines, it was taken with an adjusted angle of view so that the resolution at the horizontal edge of the CZP is 2350 TV lines, as shown in Figure 6.9a; an expanded picture is shown in Figure 6.9b. Despite no OLPF, the false signal due to aliasing at the Nyquist frequency is only slightly observed in the especially emphasized image. While it is observed in the amplitude distribution, the level is quite light compared with that of the 4.1 µm pixel in Figure 6.7. Thus, it seems that the Nyquist frequency has moved to a higher-frequency region where the MTF of the lens is not high, resulting from the achievement of higher-resolution sensors based on the progress of pixel shrinkage technology. Since the level of the false signal is light, it tends to be processed using a digital signal processor (DSP) without an OLPF. Although it depends on the application, two OLPFs are necessary for vertical and transversal directions or more than two for diagonal directions. Because one OLPF works for only one direction, it is necessary for the number of OLPF plates to be in accordance with the number of directions. Additionally, as the thickness of an OLPF is in the order of hundreds of micrometers, it is effective in achieving thinner imaging systems, thereby avoiding OLPF usage, since the trend is to reduce the size of the system.

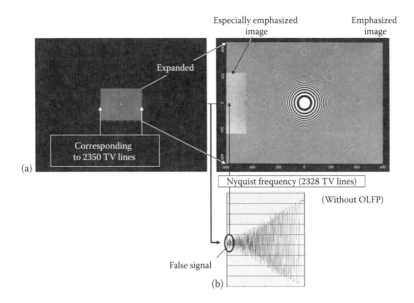

FIGURE 6.9
(a,b) CZP picture taken at 1.8 μm with a 3096(H) × 2328(V) pixel CCD without OLPF.

6.3 Sampling in Time Domain

In this section, sampling in the time domain is described. Since information concerning time is only image blurring in the case of still images, only cases of moving pictures are considered here. As already mentioned in Chapter 1, still images are repeatedly taken at a constant time interval in capturing moving images. As described in Chapter 4 on electronic shutters, images are picked up during some part of the interval time, that is, the exposure period.

This event is shown in Figure 6.10 as a schematic diagram along a time axis by the repetition of a frame time involving the exposure period. This sampling structure is exactly the same as that of the space sampling in Figure 6.4. Therefore, a shorter sampling pitch means a higher frame rate that can take higher-frequency information, that is, more accurate high-speed images can be obtained. A shorter exposure period means a narrower sampling width providing less blurred images of moving objects, that is, a higher MTF or

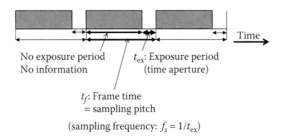

FIGURE 6.10
Schematic diagram of sampling in time domain, frame time pitch, and exposure period.

amplitude as well as space sampling. False signals exist due to aliasing during sampling in the time domain as well as space sampling. For example, on TV, the phenomenon of a rotating wheel appearing not to rotate or, inversely, appearing to rotate slowly is caused by this mechanism based on the synchronization of periodic motion and exposure timing.

6.4 Sampling in Wavelength Domain and Color Information

If the same approach as in the case of space and temporal information is adopted to obtain wavelength information, sampling by dividing the wavelength domain is considered. For higher-quality wavelength information, a higher sampling frequency and a narrower sampling width are required. Specifically, multiband cameras* capture images at each divided wavelength region, as shown in Figure 6.11, and synthesize them.

Figure 6.11 shows the case of 16 bands as the wavelength area is divided into 16 parts. In actual methods, 16 color filters, each with a spectral response that corresponds to each divided wavelength area, are prepared, and 16 still pictures are captured by using each color filter. Then, 16 pictures are synthesized to 1 color still picture. From the procedure, it is clearly understood that this system can only apply to still objects. Therefore, its application is restricted to particular kinds of tasks, such as digital archive development of art objects. And for practical reasons, the number of bands range from around 4 to 8.[†] Thus, a multiband camera system is inadequate for general application.

Almost all of the camera systems that are actually used are single-chip color cameras represented by the Bayer color filter array shown in Figure 1.7. For high-quality imaging, such as for broadcasting, professional, and high-end consumer use, the three-chip color camera[‡] is used, in which the wavelength region is separated into three parts by a prism, leading to the use of each corresponding sensor.

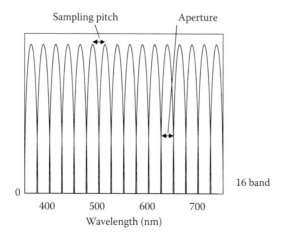

FIGURE 6.11
Example of wavelength region division for wavelength sampling.

---

* Discussed in more detail in Section 7.4.2.
† Section 7.4.2 contains an example of a 16-band camera.
‡ Discussed in more detail in Section 8.3.

An example of the spectral response of red, green, and blue in the Bayer color filter is shown in Figure 6.12. If it were thought of as one of a sampling means, it is quite different from that of space and time because of the very small sampling point number of only three and the very wide sampling widths that overlap each other. The reason is that this method in not a kind of sampling, but a method utilizing the human eye and brain's perception of color. Here, the color perception of the human eye is briefly mentioned. As is well known, the human eye processes light through the retina. The retina contains two kinds of photoreceptor cells: rods and cones. Rods only detect light intensity at a very low light level, but do not sense color. Readers might have experienced the ability to discern the shape of a body but not color under very low illumination. On the other hand, cones detect both light intensity and color in relatively bright illuminance. There are three types of cones, referred to as S, M, and L after their size.

Figure 6.13 shows the wavelength dependence of the cones' spectral response[3]: the highly sensitive wavelength ranges of S, M, and L are 400–500, 500–600, and 550–650 nm, respectively. In other words, S, M, and L are sensors that detect the ranges of violet to blue, green to orange, and yellow-green to red, respectively. Because of the overlap of the highly sensitive range, each of the three types of cones is excited by the incidence of any wavelength in visible light and responds to generate a reference stimulus.

A set of reference color stimuli generated by each of the three types of cones by light absorption facilitates color perception in the human brain. For example, if yellow, whose wavelength is around 580 nm, comes into focus at the retina, the stimulus occurs at the same level in cones L and M and at a lower level in cone S. This stimuli signal is transmitted to the brain, facilitating color perception as yellow. However, if the same intensity of green and red light is focused at the same point on the retina at the same time, the same level of stimuli occurs at cones M and L, which is transmitted to the brain and facilitates

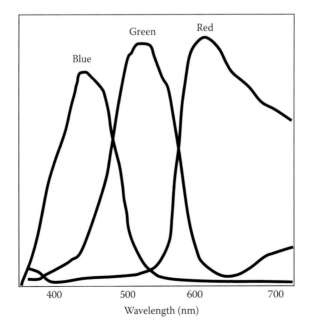

FIGURE 6.12
Example of spectral response of primary color filter for Bayer array.

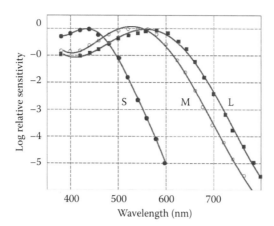

FIGURE 6.13
Relative sensitivity of cones. (Reprinted with permission from Wandell, B.A., *Foundations of Vision*, Sinauer Associates, Inc., 1995.)

the color perception. This is the the same as in the case of yellow incident light, as the brain accepts only the stimuli. Therefore, the brain perceives yellow in both cases.

When colors A and B are focused at the same point on the retina, the human color vision senses a different color C. It is not sensed as a chord-like sound. The human eye and brain detect a stimulus of a wavelength and sense as a color based on a set of stimuli at cones S, M, and L. Therefore, overlapping of the spectral response is necessary to reproduce a hue.

Display devices made up of only three primary colors can depicted as a wide range of colors because of the mechanisms of the human eye and brain.

Actually used systems are real time, such as the single-sensor camera equipped with color filters of primary or complementary colors, and the three-sensor camera. Thus, what occurs in nature is "wavelength," not "color." Color cannot be discussed exclusive of human perception. At the stage that the parameter that is a physical quantity, "wavelength," is substituted by human perception, "color," physically objective affirmation of information accuracy becomes impossible. Indeed, the colors used in filters are not completely the same, such as the case of red, green, and blue based on the primary colors and the other case of cyan, magenta, yellow, and green based on the complementary colors[4]. Various kinds of light sources can be captured, such as natural sunlight, fluorescent light, incandescent light, and light-emitting diodes (LEDs). For these reasons, there is great difficulty in physically reproducing precise color. Therefore, what is aspired to is inevitably subjective color reproduction such as perceptually equivalent color, memory color, and preferred color.

The color information of the images obtained by the method utilizing human perception has limited effectiveness for applications that human eyes do not view.

Thus, a precision signal ratio of R, G, and B is necessary for precision color information. A less-accurate ratio signal causes color error at the pixel level. Figure 6.14 shows the impact on color error caused by a random noise change according to the light intensity. The light volume increases from left to right. The signal electron numbers at the highlighted parts (forehead of doll) are 90, 180, and 300 electrons, respectively. The bottom images are expanded portions of the darker areas of the above images. While strong color error is seen in the image of 90 signal electrons, the impact of the color error decreases by signal-to-noise ratio (SNR) improvement with increasing illuminance.

Signal electron number at highlighted part

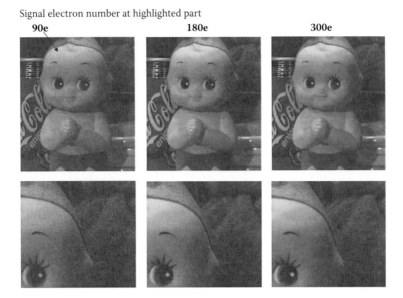

FIGURE 6.14   (See color insert)
Example of impact of random noise on color image.

Since space, color, and time are the coordinate points built in $\langle r, c, t \rangle$ space, no margin of noise occurs as a coordinate point itself. Therefore, error and defection of information of the coordinate point occur in light intensity S at the coordinate point as a false signal, such as moiré, spurious resolution, false color, color noise, blurring, and lag.

References

1. S. Nagahara, Y. Kobayashi, S. Nobutoki, T. Takagi, Development of a single pickup tube color television camera by frequency multiplexing, *Journal of the Institute of Television Engineers of Japan*, 26(2), 104–110, 1972.
2. S. Nagahara, http://www.ieice-hbkb.org/files/08/08gun_04hen_02-04.pdf, pp. 116–118, October 25, 2011 (accessed January 10, 2014).
3. B. Wandell, Foundations of color vision: Retina and brain, in *ISSCC 2006 Imaging Forum—Color Imaging*, pp. 1–23, February 9, San Francisco, CA, 2006, http://white.stanford.edu/~brian/papers/ise/ISSCC-2006-Wandell-ColorForum.pdf (accessed January 10, 2014).
4. Y. Sone, K. Ishikawa, S. Hashimoto, T. Kuroda, Y. Ohkubo, A single chip CCD color camera system using field integration mode, *Journal of the Institute of Television Engineers of Japan*, 37(10), 855–862, 1983.

7

Technologies to Improve Image Information Quality

In this book, it is explained that image information is composed of light intensity, space, wavelength, and time, and it is the role of imaging to obtain each factor with a sufficient level of information quality for the goal of the imaging system. Since imaging systems are used for various applications, the information provided is not always the same. As the most important information varies according to the purpose of the imaging system, technologies to improve performance of the important factors have been developed. In this chapter, some examples that advance the information quality of each factor are described.

7.1 Light Intensity Information

Light intensity is the most important information and contains both sensitivity and dynamic range (DR).

7.1.1 Sensitivity

Sensors measure the amount of light coming to each built-in coordinate point, and sensitivity is important to their performance. Although sensors have only been talked about thus far in terms of a sensor chip, they are actually mounted in imaging systems after they are bonded in packages sealed with transparent glass, as shown in Figure 7.1a.

Therefore, losses are incurred at the light phase before arriving at a silicon surface and at the stage of light and signal charge after penetration into the silicon. This is explained in stages. (1) Incident light decreases by about 5%–10% by reflection at both surfaces of a sealing glass, as shown in Figure 7.1a. However, sometimes antireflection coatings are used to recover the loss depending on the application, but these are generally not used due to their cost. (2) Absorption in the sealing glass is negligible except in the ultraviolet (UV) region. (3) Arriving at the sensors, part of the light is reflected at the surface of an on-chip lens (OCL). Sometimes, OCLs are covered with material having a low refractive index to suppress the loss. (4) Further, an OCL and an on-chip color filter (OCF) absorb part of the light. Since the OCF has the role of restricting the wavelength region of light that passes through it, and the spectral response is directly related to color performance, it is necessary for the OCF to absorb light that should not be transmitted through the filter of the color. (5) Absorption and reflection by a passivation film and an interlayer isolation film follow. (6) Arriving at the photodiode (PD), there is reflection at the silicon surface. A mirror-polished silicon wafer surface shows gross* reflectance in the visual region of 30%–40%, similar to metal. As this impact is never low, an antireflection (AR) film often forms, as mentioned in Section 5.1.2. (7) Before then, light that comes to the area outside the OCL and is not led to the aperture area is also lost.

---

* However, the reflection is not due to a free carrier.

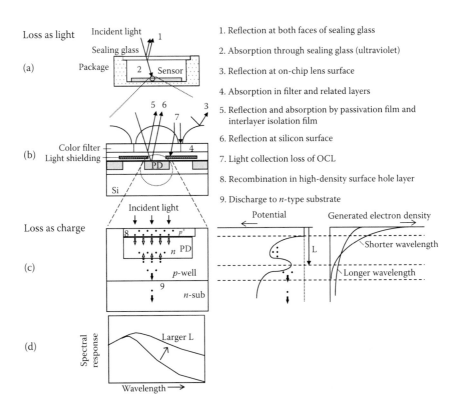

FIGURE 7.1
Loss factor of sensitivity: (a) packaged sensor; (b) sensor chip; (c) around PD; (d) diagram of *p*-well depth dependency of spectral response.

(8) On accessing the silicon, while light absorption starts photoelectric conversion, generated electrons are prone to annihilation by the recombination with high-density holes in the *p* layer near the surface, as shown in Figure 7.1c. From the viewpoint of dark current suppression, a high impurity concentration of the surface p⁺ layer is desirable. Conversely, from the viewpoint of repression of the signal charge extinction by recombination, it is preferable that the *p⁺* layer has low density and is thin. Therefore, a balanced design is necessary. (9) In the case of a sensor formed in a *p*-well on an *n*-type substrate as shown in Figure 7.1c, the signal charges generated in the *n*-type substrate are discharged. As the longer depth L, which is the distance from the surface to the electronic dividing ridge, can collect more signal charges generated in deeper areas, the sensitivity of longer wavelengths increases, as mentioned in Section 2.2.2. In the case of sensors formed on a *p*-type substrate, all the generated signal charges are capable of contributing, as shown in Section 2.2.1.

The ratio of the number of signal charges, which contribute to sensitivity, to the number of incident photons is called quantum efficiency (QE). Specifically, the ratio of the number of incident photons to the image area or a pixel is called external QE, while the ratio of the number of penetrating photons to silicon is called internal QE.

As shown in Figure 7.1, while there are many factors that impact on sensitivity, the more cost-effective technologies have been adopted. The development of technologies that make effective use of photons and signal charges, such as OCL proposed in the early 1980s, AR film, backside illumination (BSI), and advanced front-side illumination

(FSI), is ongoing, as mentioned in Chapter 5. Noise reduction technologies, which are important for signal-to-noise ratio (SNR) as sensitivity, are also still being developed, as described in Chapter 5.

7.1.2 Dynamic Range

In this section, the DR, which is the range of light intensity information that a sensor can capture, will be discussed. The DR is defined as the ratio of a signal electron number at saturation level to that of dark noise, as shown in Figure 3.6. The DR is also defined as the ratio of the maximum illuminance at which image information can be obtained without saturation to the illuminance at which the SNR equals unity. In a linear system, since a signal electron number is proportional to light intensity, both definitions of DRs are in agreement. Because it is difficult to greatly improve the DR in a linear system using state-of-the-art technology, a nonlinear system such as logarithmic conversion of photo-current[1] or a combination of multiple images information is often employed. Nonlinear systems have issues such as image lag, SNR, and temperature characteristics along with complicated signal processing according to its applications, especially color applications. Low-light performance should never be sacrificed for the sake of obtaining information on highly illuminated objects.

The following sections discuss some examples of techniques that improve the DR. Additionally, the pulse output sensor described in Section 5.3.3.2.3 is one of the methods used to increase the DR.

7.1.2.1 Hyper-D CCD

Hyper-D CCD,[2] which was proposed in 1995, can transfer twice the conventional type of signal charge packet number by forming double density VCCD, as shown in Figure 7.2, to handle a short exposure signal along with a normal one. Saturated signals in normal

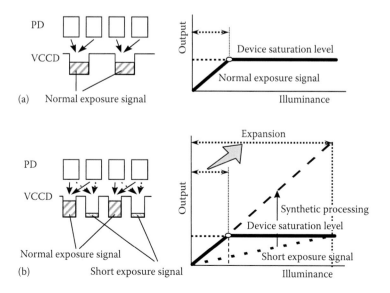

FIGURE 7.2
Conceptual diagram of hyper-D CCD: (a) conventional CCD; (b) hyper-D CCD.

exposure are replaced by nonsaturated signals in short exposure, and they are synthesized by signal processing. Thus, image information can be obtained under much higher illumination than in conventional CCDs. Examples of a captured image are shown in Figure 7.3. While the darker scene cannot be captured as well under the same exposure conditions as the highlighted scene using a conventional CCD camera, both areas are captured well by a hyper-D CCD camera. Using this signal processing, the linear relationship between light intensity and image signal as a whole image is not retained. This case is a successful example of flexibility, capability, and high functionality owing to the combination of the device with the electronic system, which was not possible in earlier film camera systems.

7.1.2.2 CMOS Image Sensor with Lateral Overflow Capacitor

The operational principle of this device[3,4] is explained in Figure 7.4. As shown in Figure 7.4a, one field-effect transistor (FET) (M3) and one capacitor (CS) are added to a four-transistor (4-Tr) pixel configuration composed of a PD, readout transistor (M1), reset transistor (M2), amplify (or drive) transistor (M4), and raw select transistor (M5). The operation is shown in Figure 7.4b. At time t1, by making M2 and M3 on-state, floating diffusion (FD) and column

FIGURE 7.3 (See color insert)
Examples of a captured dynamic range: (a) conventional CCD; (b) hyper-D CCD.

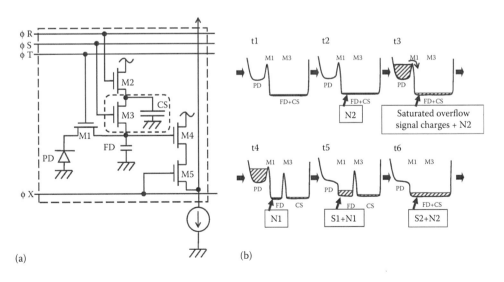

FIGURE 7.4
Wide dynamic range CMOS image sensor with lateral overflow capacitor: (a) pixel configuration; (b) schematic diagram of operational principle. (Reprinted with permission from Akahane, N., Sugawa, S., Adachi, S., Mori, K., Ishiuchi, T., and Mizobuchi, K., *IEEE Journal of Solid-State Circuits*, 41, 851–856, 2006.)

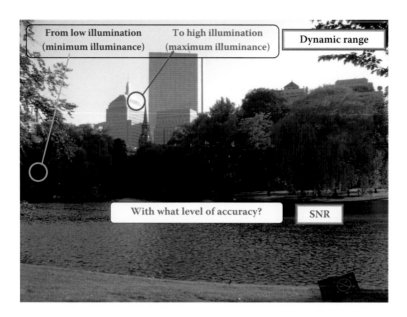

FIGURE 1.3
Example of the quality of light intensity information.

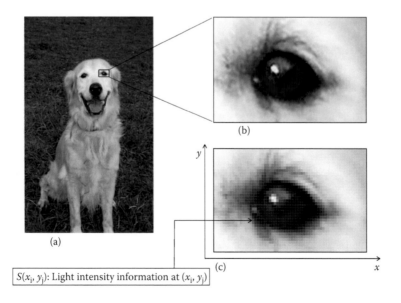

$S(x_i, y_j)$: Light intensity information at (x_i, y_j)

FIGURE 1.5
Optical image and the image signal obtained by image sensors: (a) optical image; (b) enlarged optical image; (c) image signal obtained by sensors.

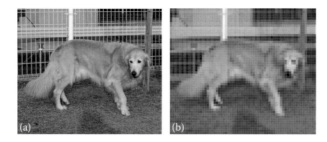

FIGURE 6.1
Comparison of sampling frequency dependence of spatial information quality: (a) fine sampling with 1318 × 1106 pixels; (b) course sampling with 64 × 54 pixels (without smoothing).

Signal electron number at highlighted part

90e   180e   300e

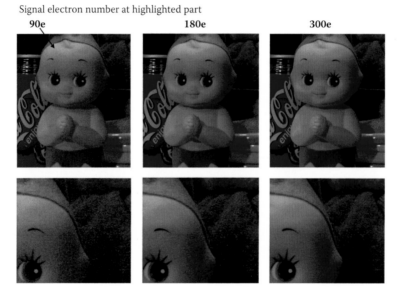

FIGURE 6.14
Example of impact of random noise on color image.

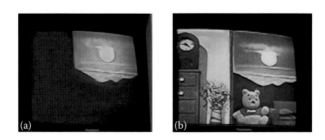

FIGURE 7.3
Examples of a captured dynamic range: (a) conventional CCD; (b) hyper-D CCD.

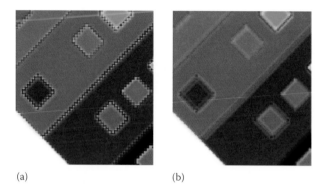

(a)   (b)

FIGURE 8.7
Example of comparison of demosaicking: (a) simple processing: shaggy at color border; (b) adaptive processing: not shaggy at color border.

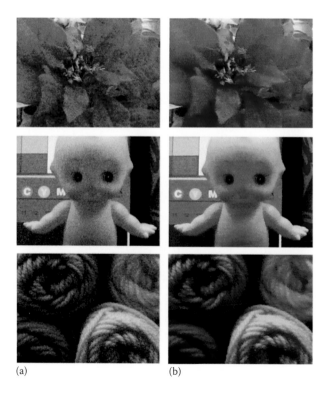

(a)   (b)

FIGURE 8.9
Examples of noise reduction by DSP: (a) original image; (b) noise-reduced image.

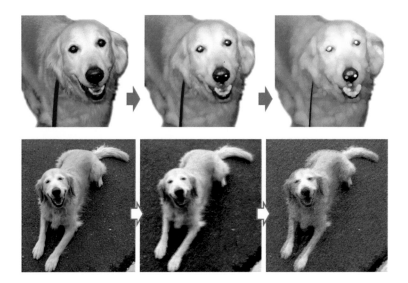

FIGURE 8.10
Problem of DSP correction dependence syndrome. Where is the border between photography and computer graphics?

select (CS) are reset to the voltage source to start the exposure. After the reset operation, the output of noise N2 existing in the combined capacitor [FD + CS] follows and is stored in off-chip memory at time t2. During the exposure period, M3 is kept on-state so that any existing oversaturated signal charges overflowing from the PD due to high illuminance are stored in the FD and CS, as shown at time t3. Before applying the readout signal charges from the PD, M3 is set to off-state. At this time, part of the oversaturated charges and noise N2 existing in the FD are output as noise N1 at time t4. Readout of the nonsaturated signal charge S1 from the PD to FD follows to output (S1 + N1) at time t5. The output voltage corresponding to S1 is obtained as the difference between the output voltages of N1 and (S1 + N1). Next, M3 is set to on-state to sum (oversaturated signal charges + N2) at time t3 and signal charge S1 to get (S2 + N2), that is, the summation of all signal charges and the initial noise N2. The summed charges amount is output by using (FD + CS) as the charge quantity detective capacitor at time t6, and the output voltage corresponding to the total signal charge S2 is obtained as the difference from the output voltage of noise N2 stored in off-chip memory. Because the noise charge quantity N2 at time t6 is the summation of N2 at time t2 and dark current generated at FD during the exposure period, they are not the same as N2 at time t2. While frame memory is necessary to store each N2 of each pixel at time t2, there is a proposal to substitute N2 by that of the next frame to avoid memory installation. While reset noise cannot be canceled because there is no correlation between the different reset operations in this case, since the signal level is higher in the oversaturated situation, it can be thought to be highly tolerant for noise.

7.2 Space Information

The improvement in space (position) information is nothing less than progress in space resolution. The most direct method to enhance the Nyquist frequency is to increase the pixel number. The pixel interpolation array, described in Section 5.2.3.1, increased the horizontal resolution by devising a pixel array without increasing the pixel number. This sensor was produced for video cameras in the early 1980s. Because the scanning line number is decided by the format of television systems, there was no need to increase the vertical resolution. A sensor that also extends to vertical resolution in the same manner as for digital camera use is a pixel interleaved array CCD (PIA CCD).[5]

Square and interleaved pixel arrays in real space are shown in Figure 7.5a and b, with the interleaved array indicating a rotation at an angle of 45° of the square array. The vertical, horizontal, and diagonal pixel pitches in the square array are p, p, and $p/\sqrt{2}$, respectively. Conversely, the vertical and horizontal pitches in the interleaved array are shortened to $p/\sqrt{2}$, while the diagonal pitch is p as shown in Figure 7.5b.

The Nyquist frequency obtained by Equation 6.1 is shown as frequency space in Figure 7.5c. In an interleaved array, the Nyquist frequency is higher than that of a square array in the vertical and horizontal directions, while that of the diagonal direction is lower. Thus, some part of a higher resolution in the diagonal direction of a square array is allocated to the vertical and horizontal directions in the interleaved array. Since the configuration is only rotated at an angle of 45°, the sampling density, that is, the information density, is the same, but the weight is changed in accordance with the directions.

So is there much point in it? The answer is "yes." Watanabe et al.[6] report that the human eye has higher sensitivity in vertical and horizontal directions. Additionally, in a paper on PIA,[5]

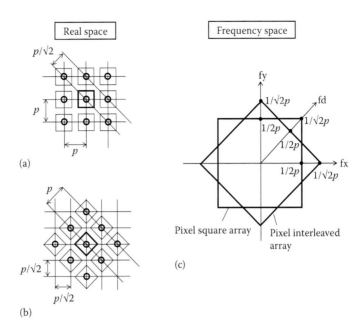

FIGURE 7.5
Pixel arrays in real space and Nyquist frequency of each array: (a) pixel square array; (b) pixel interleaved array; (c) Nyquist frequency.

it was shown that the structure ratio of vertical and horizontal lines are statistically higher than other angles in huge number of images. Thus, a way to create efficiency in accordance with the characteristics of the human eye and the statistical nature of objects is desirable.

The explanation so far is for the case of monochromatic use, in which each pixel has correlations with its neighboring pixels and a reasonable response or output can be supposed from each pixel. But the situation is quite different in the case of a one-chip color system, that is, one color filter at one pixel.

Before the discussion proceeds, some points concerning the characteristics of the human eye should be made. As mentioned in Section 6.4, the human eye and brain perceive color as a set of stimulus values of cones S, M, and L. There, what contributes most, that is, the highest sensitivity frequency is around 550 nm, which is from green to yellow-green. So, the pixel that contributes most to sensitivity and resolution is the green filter pixel; therefore, the green pixel array is important.

First, the Bayer configuration filter applied to a pixel square array sensor is discussed. Each Nyquist frequency of G, R, and B is considered as well as the previous case of monochrome by focusing on the array of each color.

As shown in Figure 7.6a, the G pixel configuration is an interpolated array and the vertical and horizontal pitch is p, which is the same as in the monochrome case. Therefore, the Nyquist frequency of green in the vertical and horizontal directions is $1/2p$, which is the same with monochrome as shown in Figure 7.6b, which in the monochrome case is indicated by B/W. Thus, the combination of a square array sensor and the Bayer configuration is well made.* The pitches of R and B in the vertical and horizontal directions are $2p$ and the Nyquist frequency is $1/4p$, as shown in the figure.

---

* This is not surprising because the concept of Bayer's invention (Reference [6] in Chapter 1) is to arrange the resolution contributory color in a checkerwise fashion.

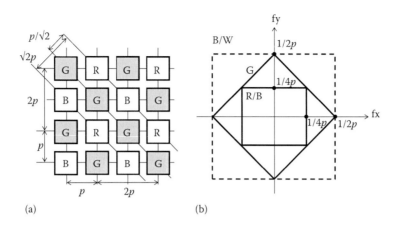

(a)   (b)

FIGURE 7.6
(a) Pixel square array with Bayer configuration color filter; (b) Nyquist frequency of each color.

Conversely, in the combination of an interleaved pixel array sensor and the Bayer configuration shown in Figure 7.7a, the G pixel configuration is a square array with the pitches of $\sqrt{2}p$ in the vertical and horizontal directions; therefore, the Nyquist frequency is $1/2\sqrt{2}p$, as shown in Figure 7.7b, indicating debasement to one-half of the monochrome case and lower than the square pixel array sensor, while the diagonal direction is as high as $1/2p$. As the pitch of R and B colors in the vertical and horizontal directions is $\sqrt{2}p$, the same as G, the Nyquist frequency is also $1/2\sqrt{2}p$.

This is also one of the technologies that, in principle, has an advantage with straightforward validity for monochrome image capture, but is quite different for color image capture by single-chip color systems.

Since many objects in near achromatic color overlap the spectral response of the green color filter with that of the red and blue filters, decay at this level is infrequent. But a rather complicated signal processing is required as well as a wider overlapping of the spectral response between colors, which tend to degrade the hue accuracy. Because an interleaved array

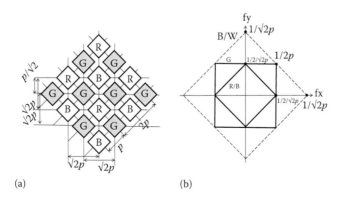

(a)   (b)

FIGURE 7.7
(a) Pixel interleaved array with Bayer configuration color filter; (b) Nyquist frequency of each color. (Reprinted with permission from Kosonocky, W., Yang, G., Ye, C., Kabra, R., Xie, L., Lawrence, J., Mastrocolla, V., Shallcross, F., and Patel, V., *Proceedings of the IEEE Solid-State Circuits Conference Digest of Technical Papers*, 11.3, pp. 182–183, San Francisco, CA, 1996.)

configuration is essentially on the basis that every pixel has a proper level of response, many combinations are required at various points, such as each color spectral response, color configuration, algorithm for color, and luminance signal processing. Therefore, depending on the color distribution of the objects, the effect does not work, or, adversely, it falls to a low level.[7]

The ratio of the pixel numbers of R, G, and B is 1:2:1 in the Bayer configuration; however, there is an example of improvements in resolution and sensitivity by increasing the number of G in the ratio to 1:6:1 for camcorder applications.[8]

7.3 Time Information

The improvement of time information is essentially high time resolution.

7.3.1 Frame-Based Sensors

In the frame-based integration mode in which exposure is prosecuted with a predefined periodicity and exposure period, which almost all image sensors employ, a higher time resolution and a higher modulation transfer function (MTF) can be obtained using a higher frame rate and a shorter exposure period, as mentioned in Section 6.3. To that end, the task is to find how a higher frame rate (frames/s: fps) or a higher output pixel number rate (pixels/s) can be realized. The expedients for frame-based sensors are as follows:

1. Parallel output by multiple channels
2. High-speed digital output by column-parallel analog-to-digital converter (ADC)
3. Burst-type sensor with the required number of built-in frame memories

7.3.1.1 Parallel Output–Type Sensor

In this method, because the pixel signal output is shared by parallel multiple output channels, the output frequency can be increased by the number of output channels, as shown in the conceptual diagrams in Figure 7.8a and b, illustrating a 4-channel output CCD and an

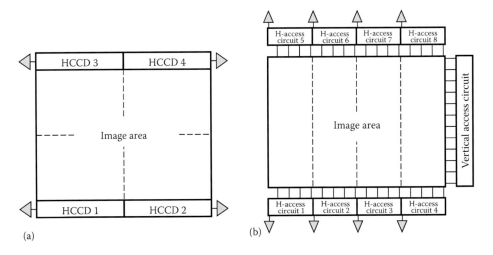

FIGURE 7.8
Examples of parallel output sensors: (a) CCD; (b) CMOS sensor.

8-channel output CMOS sensor, respectively. There is an example of a CMOS sensor with 128 channel parallel outputs[9] by 64 channel outputs at the top and bottom, respectively. It should be considered that fixed-pattern noise is apt to be caused as output differences due to the variation of the output channel characteristics independent of the sensor type, and countermeasures are necessary.

7.3.1.2 Column-Parallel ADC-Type Sensor

This type of sensor is described in detail in Section 5.3.3.2.2. As mentioned there, various types of column converter ADC sensors are reported and a high-rate output performance of gigapixels per second is achieved. Continuous progress is expected in accordance with its applications.

7.3.1.3 Burst-Type Sensor

All the sensors described thus far have continuous output with continuous capturing. In contrast, in a burst-type sensor, on-chip frame memories are equipped to store captured image signal charges for giving priority to higher frame rate by avoiding output operation at each frame, which needs driving and output time. The first one is burst CCD,[10] proposed in 1996.

A two-by-two pixel configuration is shown in Figure 7.9; signal charges generated in the PD are collected and stored in the potential well under the G1 gate. They are transferred in series into the serial-parallel register of the pixel to detect successive frames by way of the G3 gate channel. The signal charges of the following frame are also transferred into the serial-parallel register. After the serial-parallel register is filled with the signals of five frames, the charge signals are transferred in parallel from the serial-parallel register to the parallel

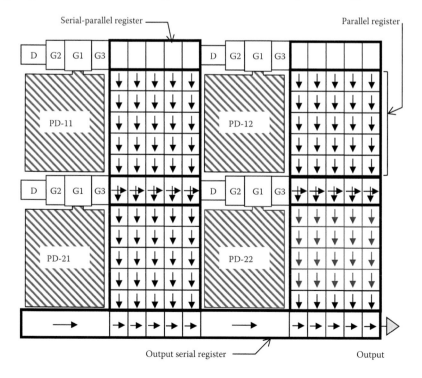

FIGURE 7.9
Pixel configuration of burst-type CCD (serial-parallel memory).

register. By repeating this action, five-by-six frame memories are filled. This operation is continued until the desired phenomenon is observed. During the operation, when the signal charges are transferred in parallel from the serial-parallel register to the parallel register, five signal charge packets in the last row in the parallel register of the pixel above are transferred to the serial-parallel register of the lower pixel. At each transfer of new signal charges from the PD to the serial parallel register in series, five signal charges transferred from the above pixel memory are transferred to the dumping drain D in series too.

Since this sensor has a register that can store 30 signal charge packets at each pixel, the number of frames of images that can be captured is 30. This is an issue of the range of time information on the accuracy and range of the four factors mentioned in Section 1.1.

While this sensor was aiming to realize 10^6 fps, the accomplished frame rate was 3×10^5 fps at the time of presentation at the conference. The *in situ* storage image sensor (ISIS)[12] was devised following the advice that linear CCD-type memory should be employed to pursue a higher frame rate.[11] This advice was based on an insight that the serial-parallel register of the above sensor made this goal difficult to achieve, an idea almost impossible to conceive for those who have never actually developed CCDs.

As shown in Figure 7.10, a linear CCD-type memory with 103 stages is formed at each pixel and 10^6 fps is achieved. Signal charges under a photogate sensor are transferred to the linear CCD memory at each 1 μs, and signal charges in the memory are transferred at the same time, the head charge packet arriving at the drain is discharged. This operation is repeated until the desired phenomenon is observed. In 2011, this sensor was improved

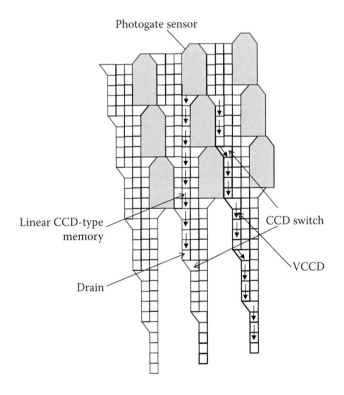

FIGURE 7.10
Pixel configuration of ISIS (linear CCD-type frame memory). (Reprinted with permission from Etoh, T., Poggemann, D., Ruckelshausen, A., Theuwissen, A., Kreider, G., Folkerts, H., Mutoh, H. et al., *Proceedings of IEEE International Solid-State Circuits Conference Digest of Technical Papers*, 2.7, pp. 46–47, San Francisco, CA, 2002.)

on[13] to realize a frame rate as high as 1.6×10^7 fps by using BSI for high sensitivity and charge multiplication. However, it does have issues, such as increased load on cooling of the system to suppress the rise in heat caused by high-frequency driving of a CCD, which is a large capacitor.

7.3.1.4 Coexistence Type of Burst and Continuous Imaging Modes

A burst-type CMOS image sensor was also developed. By putting the CMOS sensor to good use, analog frame memory areas are formed separate from the image area in the sensor,[14] as shown in Figure 7.11 with $400(H) \times 256(V)$ pixels.

On the top and bottom of the image area, a frame memory of 128 frames/pixel is formed for temporary storage. Signals are read out from the image area to the memory at high speed through 32 signal lines in each column, that is, four pixels per one output. The sample and hold circuit of a CDS arranged as internal circuits in a pixel provide global shutter function. Interestingly, it can capture both the burst imaging mode of 10^7 fps (1 Tpixels/s) with 128 frames of 10^5 pixel numbers and the continuous imaging mode of

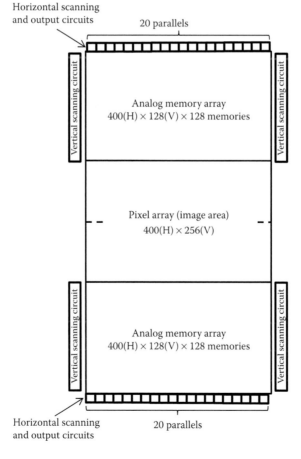

FIGURE 7.11
Architecture of high-speed CMOS sensor. (Reprinted with permission from Tochigi, Y., Hanzawa, K., Kato, Y., Kuroda, R., Mutoh, H., Hirose, R., Tominaga, H., Takubo, K., Kondo, Y., Sugawa, S., *Proceedings of IEEE International Solid-State Circuits Conference Digest of Technical Papers*, 22.2, pp. 382–384, San Francisco, CA, 2012.)

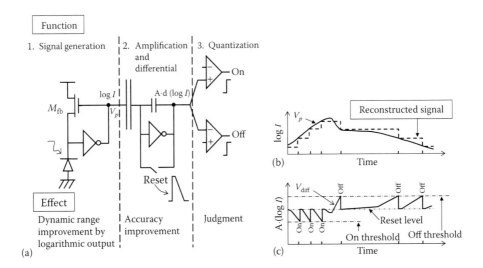

FIGURE 7.12
Event-driven sensor: (a) pixel schematic and the role of each block; (b) log I and reconstructed log signals; (c) temporal transition of A·d(log I). (Reprinted with permission from Lichtsteiner, P., Posch, C., Delbruck, T., *Journal of Solid-State Circuits*, 43, 566–576, 2008.)

7.8 k fps (780 Mpixels/s) of 10^5 pixel numbers with analog parallel output. When additional measures are included to achieve high speeds, it is an impressive imager filled with ideas.

7.3.2 Event-Driven Sensor

The sensor described here is the second example that does not belong to "almost all image sensors"[15] in this book. It is not a frame-based type of sensor in which signal charges are integrated for a prefixed exposure period. Its principle of operation is completely different except for photoelectron conversion.

As shown in Figure 7.12a, a pixel is composed of three blocks.[16] In (1), the signal generation block, a photocurrent through a PD is not integrated but monitored constantly. The voltage at the node, which is connected to the cathode of the PD and the source of the feedback transistor, M_{fb}, is amplified by an inverting amplifier, and the output is connected with the gate input of M_{fb} to form an amplified and log-transformed voltage output of photocurrent I. The output is transferred to (2) the amplification and differential block, and is temporally differentiated; however, only the component that varies with time is amplified. The output is transmitted to (3), the quantization block, and is monitored by an increase/decrease checking comparator, which emits a pulse signal of ON or OFF in accordance with an increase or decrease in input change, when a predefined amount is observed. The signal is generated by pulses indicating an increase or decrease in the predefined amount of input change only at the time of change and only at the changed pixel, there is no signal showing light intensity directly. In that context, this sensor has communality* with the pulse output sensor described in Section 5.3.3.2.3. Therefore, light intensity or the change of light intensity information is quantized in these sensors differently from "almost all sensors," while time information is not quantized.

Thus, the obtained information is only the pixel address and time, and the increase or decrease at which the predefined amount of change occurred. That signals are generated

---

* Neither of these sensors belongs to the frame-based integration mode type, but both emit a pulse signal and reset monitoring integration to restart, when the light intensity information changes to a predefined amount.

only at pixels and at times when the input has changed is a big feature of this sensor, and is quite different from frame-based sensors. Therefore, redundancy is significantly reduced compared with the frame-based mode in which signals are read out at each pixel at each frame regardless of whether input changes. The generated signal pulses in the quantization block are also used for the reset operation in the second block, from where the next potential monitoring starts. If the output of the first block, log I or V_p, changes as shown by the solid line in Figure 7.12b, the output of the second block, A·d(log I), appears. When it reaches the threshold of ON or OFF, which are the predefined amounts of change, a signal pulse is emitted by the third block as shown in Figure 7.12c and the second block output is reset. From the obtained time information when the predefined change occurred as shown in Figure 7.12c, the signal shown by the dotted line in Figure 7.12b can be obtained as a reconstructed temporal transition of log I or V_p. In this reconstruction, while it can be said that the light intensity information is quantized by a predefined amount of change of the amplified differential signal, the time information is obtained as high as the resolution determined by circuit characteristics. Thus, the time information is not quantized in this sensor, since this is not a frame-based sensor. In this sensor, the quantized factors, or built-in coordinate points, are space and amount of change of time-derivative log-transformed voltage output of amplified photocurrent I. Therefore, the signal output of the sensor is not the amount of integrated signal charge $S(r, t)$, but time T when amount of change of time differential of log-transformed voltage output of amplified photocurrent I reaches a predetermined quantity $\pm A\cdot\Delta[\partial\log I/\partial t]q$ at pixel r_k, that is, $T(\pm A\cdot\Delta[\partial\log I/\partial t]q, r_k)$, where A is voltage gain of amplification.

The time resolution of this sensor is higher than 10 µs. The DR is 120 dB by logarithmic output and differential circuit. Because only input-varied pixels emit a signal pulse, a moving object composes a cluster of those pixels, and it can be recognized as a moving object in real time. This sensor has 26 transistors in a pixel.

Readers are directed to the website of the research institute of this sensor,[17] where various interesting moving images are shown on the homepage.

7.4 Color and Wavelength Information

The physical quantity named "wavelength" exists in the natural world and "color" is a perception generated by the human eye and brain. As mentioned in Section 6.4, since it is quite difficult to obtain images with physically accurate wavelength information using the current technology, a subjective color reproduction technique is commonly used for applications that are viewed by the human eye. In this field, a single-sensor color camera system represented by the primary colors R, G, and B or the complementary colors magenta, yellow, cyan, and green and the three-sensor camera system with the primary colors are used.

7.4.1 Single-Chip Color Camera System

Color reproduction by three or four colors is an absolute approximation. Therefore, it is possible to express more subtle shades of color by adding a new color with an appropriate spectral response; however, this usually has side effects such as degradation of the SNR. As system designs are determined by what characteristics should be featured as part of a balanced overall performance, designs unsuitable for high sensitivity, which is the priority for common imaging systems, are not commonly adopted. In a digital still camera system, the Bayer configuration color filter remains dominant.

7.4.2 Multiband Camera System

While the color information of a pixel in a single-chip color system is signified by a set signal of R, G, and B, a multiband camera can obtain much more detailed color or wavelength information, as mentioned in Section 6.4.

As shown in Figure 7.13, multiple color filters attached to a turret are set in front of a photographic lens in a multiband camera system, and still images are captured through each color. It has been clarified that, in principle, 99% of color information can be obtained by three kinds of color filters.[18] However, as it is quite difficult to realize an ideal combination of color filters, a practical solution was proposed that 99% of color information can be achieved by using five kinds of commercially available pigment color filters. While this system was developed for up to 8-band cameras, there is an example of a 16-band camera.[19]

7.4.3 Hyperspectral Imaging System

A hyperspectral imaging system provides a thorough viewpoint of a multiband camera. While the objective of a multiband camera with a smaller band number is to obtain higher color information, that of a hyperspectral imaging system is precise and wide-ranging wavelength information. A hyperspectral imaging system can be considered as a camera with a spectroscopic device such as built-in grating[20] rather than a multiband camera with a restricted number of color filters. Hyperspectral imaging systems with a 5 nm bandwidth and around 100 band numbers have been developed and are considered the ultimate multiband camera.

The operational principle of hyperspectral imaging is shown in Figure 7.14. A linear portion of a whole image is passed through an optical slit and is dispersed by a spectroscopic device in a vertical direction to the incident light of the image sensor. Focusing on one vertical line on the sensor, the spectrum can be obtained from the longest wavelength at the top edge pixel to the shortest wavelength at the bottom pixel. On completion of the readout of one image on an image sensor, the linear image part is shifted to the next one to be captured in the following step and in the same manner. By scanning a two-dimensional image from the top line to the bottom line, a hyperspectral image is obtained. Since the principle is the same with a multiband camera, it is not suitable for real-time reproduction.

Because of the principle, a target is not limited to the visible region. While an optics system and sensor must be chosen, their objectives are wide from UV, visible, near-infrared, infrared, to far-infrared imaging. Among these functions, hyperspectral imaging is applied in a wide

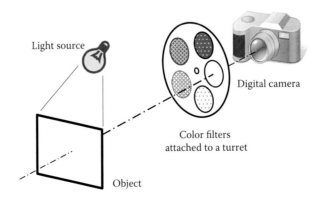

FIGURE 7.13
Schematic diagram of multiband camera.

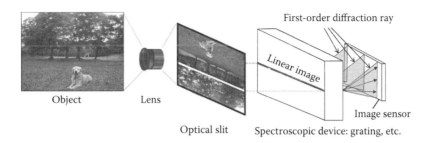

FIGURE 7.14
Operational principle of hyperspectral imaging.

range of areas such as quality review and safety of agriculture and food, medical science, biotechnology, life science, and remote sensing, and further advancement is anticipated.

References

1. S.G. Chamberlain, J.P.Y. Lee, A novel wide dynamic range silicon photodetector and linear imaging array, *Transaction on Electron Devices*, ED-31(2), 175–182, 1984.
2. H. Komobuchi, A. Fukumoto, T. Yamada, Y. Matsuda, T. Kuroda, 1/4 inch NTSC format hyper-D range IL-CCD, in *IEEE Workshop on CCDs and Advanced Image Sensors*, April 20–22, Dana Point, CA, 1995, http://www.imagesensors.org/Past%20Workshops/1995%20Workshop/1995%20 Papers/02%20Komobuchi%20et%20al.pdf (accessed January 10, 2014).
3. N. Akahane, S. Sugawa, S. Adachi, K. Mori, T. Ishiuchi, K. Mizobuchi, A sensitivity and linearity improvement of a 100 dB dynamic range CMOS image sensor using a lateral overflow integration capacitor, in *2005 Symposium on VLSI Circuits, Digest of Technical Papers*, pp. 62–65, June 16–18, Kyoto, Japan, 2005.
4. N. Akahane, S. Sugawa, S. Adachi, K. Mori, T. Ishiuchi, K. Mizobuchi, A sensitivity and linearity improvement of a 100-dB dynamic range CMOS image sensor using a lateral overflow integration capacitor, *IEEE Journal of Solid-State Circuits*, 41(4), 851–856, 2006.
5. T. Yamada, K. Ikeda, Y. Kim, H. Wakoh, T. Toma, T. Sakamoto, K. Ogawa, et al., A progressive scan CCD image sensor for DSC applications, *Journal of Solid-State Circuits*, 35(12), 2044–2054, 2000.
6. A. Watanabe, T. Mori, S. Nagata, K. Hiwatashi, Spatial sine-wave responses of the human visual system, *Vision Research*, 8, 1245–1263, 1968.
7. H. Miyahara, The picture quality improve technology for consumer video camera, *Journal of ITE*, 63(6), 731–734, 2009.
8. Sony Corporation. http://www.sony.jp/products/Consumer/handycam/PRODUCTS/special/02cmos.html (accessed January 10, 2014).
9. B. Cremers, M. Agarwal, T. Walschap, R. Singh, T. Geurts, A high speed pipelined snapshot CMOS image sensor with 6.4 Gpixel/s data rate, in *Proceedings of 2009 International Image Sensor Workshop*, p. 9, June 22–28, Bergen, Norway, 2009, http://www.imagesensors.org/Past%20Workshops/2009%20Workshop/2009%20Papers/030_paper_cremers_cypress_gs.pdf (accessed January 10, 2014).
10. W. Kosonocky, G. Yang, C. Ye, R. Kabra, L. Xie, J. Lawrence, V. Mastrocolla, F. Shallcross, V. Patel, 360×360-element very-high frame-rate burst-image sensor, in *Proceedings of the IEEE Solid-State Circuits Conference Digest of Technical Papers*, 11.3, pp. 182–183, February 8–10, San Francisco, CA, 1996.
11. T. Kuroda, Private communication at Kyoto Research Laboratory, Panasonic Corporation, April 1996.

12. T. Etoh, D. Poggemann, A. Ruckelshausen, A. Theuwissen, G. Kreider, H. Folkerts, H. Mutoh, et al., A CCD image sensor of 1 Mframes/s for continuous image capturing of 103 frames, in *Proceedings of IEEE International Solid-State Circuits Conference Digest of Technical Papers*, 2.7, pp. 46–47, February 3–7, San Francisco, CA, 2002.

13. T. Etoh, D. Nguyen, S. Dao, C. Vo, M. Tanaka, K. Takehara, T. Okinaka, et al., A 16 Mfps 165 kpixel backside-illuminated CCD, in *Proceedings of IEEE International Solid-State Circuits Conference Digest of Technical Papers*, 23.4, pp. 406–408, February 20–24, San Francisco, CA, 2011.

14. Y. Tochigi, K. Hanzawa, Y. Kato, R. Kuroda, H. Mutoh, R. Hirose, H. Tominaga, K. Takubo, Y. Kondo, S. Sugawa, A global-shutter CMOS image sensor with readout speed of 1 Tpixel/s burst and 780 Mpixel/s continuous, in *Proceedings of IEEE International Solid-State Circuits Conference Digest of Technical Papers*, 22.2, pp. 382–384, February 19–23, San Francisco, CA, 2012.

15. P. Lichtsteiner, C. Posch, T. Delbruck, A 128×128 120 dB 30 mW asynchronous vision sensor that responds to relative intensity change, in *Proceedings of IEEE International Solid-State Circuits Conference Digest of Technical Papers*, 27.9, pp. 508–510, February 6–9, San Francisco, CA, 2006.

16. P. Lichtsteiner, C. Posch, T. Delbruck, A 128×128 120 dB 15 μs latency asynchronous temporal contrast vision sensor, *Journal of Solid-State Circuits*, 43(2), 566–576, 2008.

17. T. Delbruck, Dynamic vision sensor (DVS): Asynchronous temporal contrast silicon retina, siliconretina, 2013. http://siliconretina.ini.uzh.ch/wiki/index.php (accessed January 10, 2014).

18. Y. Yokoyama, T. Hasegawa, N. Tsumura, H. Haneishi, Y. Miyake, New color management system on human perception and its application to recording and reproduction of art paintings (I)—Design of image acquisition system, *Journal of SPIJ*, 61(6), 343–355, 1998.

19. M. Yamaguchi, T. Teraji, K. Ohsawa, T. Uchiyama, H. Motomura, Y. Murakami, N. Ohyama, Color image reproduction based on the multispectral and multi-primary imaging: Experimental evaluation, *Proceedings of SPIE*, 4663, 15–26, 2002.

20. Shin, Satori. http://www.nikko-pb.co.jp/nk_comm/mok08/html/images/1203g61.pdf (accessed January 10, 2014).

8

Imaging Systems

In earlier chapters, the interior, function, and driving mechanisms of image sensors have been described. In this chapter, factors having an impact on the quality of image information and the signal processing done after sensor output will be discussed to provide an outline of whole imaging systems.

8.1 Deteriorating Elements of Image Information Quality

In imaging systems, light from self-luminous objects or objects illuminated by a light source is focused on the image sensor through an optics system such as a lens, and the optical image information is converted to an electrical image signal, as shown in Figure 8.1.

The image signal is digitized by way of an analog noise reduction circuit and processed by a digital signal processor (DSP) to make it a color image signal. Image information is composed of four factors, as mentioned in Chapter 1; the quality of the information translates directly into image quality. Elements that exert influence on the information quality of each factor at each step are considered in terms of noise (inaccuracy) and range (limit).[1]

In Figure 8.2, the elements of the light source and optics are shown. The stability of the light source is most important. In the case of light sources that come and go periodically, such as fluorescent lamps, flicker noise* makes image brightness fluctuate with time, depending on the frame frequency and exposure period. The second type of noise generated by the light source, optical shot noise, is universal noise and consists of variance of the photon number itself, as described in Section 3.4. Light reflected by the object arrives at the lens, where shading, flare, and modulation transfer function (MTF) exist as intensity noise. Space noise is composed of aberration (distortion), focusing, diffraction phenomenon caused by the iris (Airy disk), and camera shake. Chromatic aberration is wavelength noise. In an infrared cut filter, spectral response is wavelength noise. As optical low-pass filters only remove single-wavelength light, this is the origin of space noise for light of other wavelengths. The major light intensity noises in image sensors were mentioned in Chapter 3 and are detailed in Figure 8.3.

While the noises described above are involved in the light coming to the sensor, the following are generated in the sensor.

Intensity noise in the sensor is composed of electronic noise and optical noise. Electronic noise is composed of device noise including transistor noise, to which 1/f noise, random telegraph noise (RTN), and thermal noise belong, and circuit noise, to which kTC noise, vertical line noise caused by column circuit variance, noise that was not canceled enough because of insufficient correlated double sampling (CDS) efficiency, and shading noise belong. Further, sensitivity unevenness or shading caused by variation of the light focusing effect of the on-chip micro lens (OCL) is intensity noise.

The dynamic range restricts the intensity signal level that the sensor can treat.

---

* Eastern Japan is the only area where the frequencies of commercial power and television systems are different.

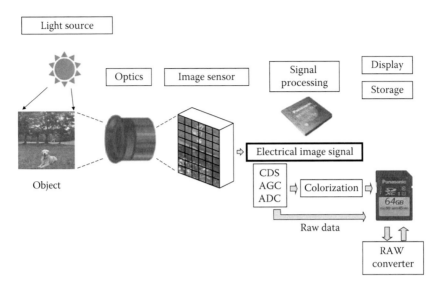

FIGURE 8.1
Overall flow of imaging system. AGC, automatic gain control.

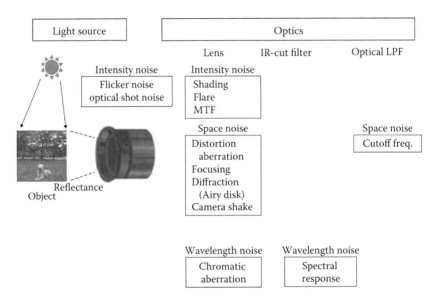

FIGURE 8.2
Image information and elements of quality: light source and optics.

Spurious resolution around the Nyquist frequency and cross talk of light or signal charge between pixels are also space noise. While the space range is determined by the area size of the image, it also depends on the focal length of the lens. Wavelength noise includes false color due to periodic space sampling, spectral response of the color filter, cross talk of light and signal charge, and overlap between color filters. However, because the spectral response of the color filter is decided by a balance of signal-to-noise ratio (SNR), color reproducibility, resolution, arithmetic load, and so on, this cannot be said as a rule. Wavelength range is the color-reproducible area and depends on the number of filter colors and the

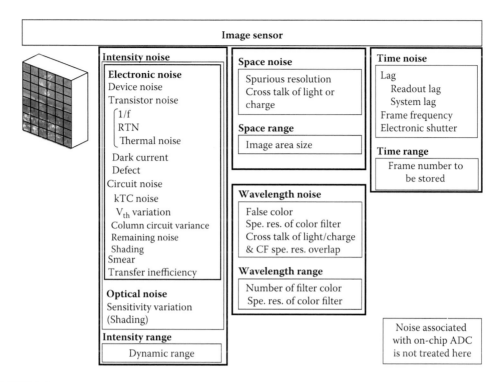

FIGURE 8.3
Image information and elements of quality: image sensor.

spectral response of the color filters in either the visible region or the region outside of the visible region, depending on the application. Time noise includes lag phenomena caused at readout from the photodiode and system lag due to the overlap of time information between successive frames, which is rarely seen these days. Aliasing due to periodical sampling, mentioned in Section 6.3, depends on the relation between frequencies of sampling and reciprocal motion of the object. Picture blur of an object moving at high speed during the exposure period, which depends on exposure time, is time noise as well as space noise. Time range is the number of frames that can be stored and is not commonly an issue, except for burst-type sensors, in which it is determined by the built-in frame memory volume.

Next, the signal processing part is considered. In the analog front end (AFE), after correlated noise involved in the analog output signal is reduced through CDS operation, the signal is amplified with adaptively chosen voltage gain in accordance with signal voltage and is translated to a digital signal through an analog-to-digital converter (ADC), as shown in Figure 8.4. Digital output sensors described in Section 5.3.3.2 complete this step.

Various operations are processed in the digital domain by a DSP. Operated processing is not restricted to what should be essentially processed for color image signal reproduction, such as demosaicking (color incorporation), color conversion (color matrix), white balance, and gamma processing, but is extended to various kinds of correction such as defect correction, noise cancellation or reduction, chromatic aberration correction and shading correction, to improve the appearance of the image.

In this way, high revision processing can be realized, which was not possible for analog processing. It could be said that the light makeup of the days of analog signals has changed to heavy makeup in this digital signal age without attracting attention. Therefore,

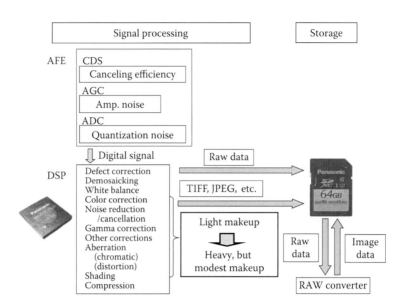

FIGURE 8.4
Image information and elements of quality: signal processing.

it is almost impossible to discuss camera performance by referring to only sensor performance without referring to functions and performances of DSPs. Even in the case of raw data output, it is quite rare for raw data to be output without any basic correction such as defect correction. Data that are finely converted to formats such as TIFF and JPEG are stored in memory.

As observed above, there are many elements that influence information quality. It is important to comprehend the degree, influence, and priority of these elements.

8.2 Signal Processing

To return to a subject mentioned in Section 8.1, what DSPs should originally process is described. Digitization of the sensor output signal alone cannot make it a color image signal the human eye can process, but various processing operations are necessary. An example outline of the overall process follows.

Each color signal, R, G, and B, from the sensor output is selected and used to build an RGB color image. The signal is then corrected to easily viewable brightness, adjusted to a natural color with enhancement of resolution and contrast, and converted to a format such as TIFF or JPEG.

An example of processing flow is shown in Figure 8.5. In the first stage, linear operation of pixel data is carried out; after corrections of defect, brightness, and white balance, demosaicking and color conversion follow. Nonlinear processing follows, such as color/tone conversion, noise reduction, and edge enhancement. A series of these types of signal processing is an important operation that has a big impact on the final image quality or impression. This is called the "image processing engine" and camera manufacturers put great effort into it.

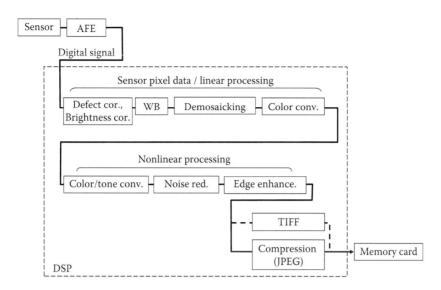

FIGURE 8.5
An example of signal processing flow.

In the following sections, each element of the processing series shown in Figure 8.5 will be described.

8.2.1 Defect Correction, Brightness Correction

While defect-free (white and black defects) sensors are desirable, sensor cost would be quite expensive if only perfect zero-defect sensors could be applied for imaging systems. Therefore, sensors having defects within the correctable range of level and number are provided for practical use. To correct the defect, the signal of the defective pixel is substituted by that of another, normal pixel or a signal obtained by peripheral normal pixels. In brightness correction, the signal level is adjusted for easy subsequent processing.

8.2.2 White Balance

Adjusting white balance (WB) prevents objects that should be expressed in white from being expressed in tints other than white using a light source. There are two ways of doing this. One is to take spectral distribution information of the light source directly from sensor output distribution and adjust WB at the time of photography. The other is to adjust WB during signal processing after image capture by color distribution of the captured signal. Since the capabilities of DSPs are now advanced, the latter method is preferable from the aspects of imaging system size, cost, time, and effort at the time of photography. In the latter method, the ratio between R, G, and B is adjusted based on the assumption that summation of pixel output of an image is achromatic (gray). But this method fails when most of the image is highly chromatic, such as a view illuminated by a sunset, because the above assumption does not match. Therefore, most cameras have other options for WB.

8.2.3 Demosaicking

As only one filter color is formed at each pixel in a one-sensor color camera system, only one piece of color information is obtained at each pixel. On the other hand, the human eye and brain perceive a color by a set of R, G, and B, as mentioned in Section 6.4. Therefore, all R, G, and B information is necessary at each pixel to construct a color image for applications for humans to watch. Demosaicking creates color signals that are not captured originally at each pixel.

As shown in Figure 8.6, pixel signals from sensor output are separated by every color, and color signals lacking at each pixel represented as R', G', or B' in Figure 8.6c are generated in the demosaicking operation for color signal interpolation. Thus, every color signal at each pixel is obtained. In other words, demosaicking creates a flawless color signal, as shown in Figure 1.10a, from a color signal obtained by a single-chip color camera system, as shown in Figure 1.10b.

Since many subtle aspects such as smoothness of boundary and color noise level are determined by the architecture of the demosaicking process, each camera company exercises their ingenuity to develop this. As a technique to create color signals that are lacking, correlation relations between colors are often used such that red signal distribution is assumed to be proportional to green distribution in low-frequency domains. But if this is executed in a simple way, conspicuous color errors are likely to be generated at color boundaries, so it is necessary to process adaptively by checking patterns in detail. As an example, adaptively processed demosaicking is shown in Figure 8.7, and a definite effect is seen in whether it is shaggy or not on the border of the color. Algorithms such as this are another source of differentiation between companies.

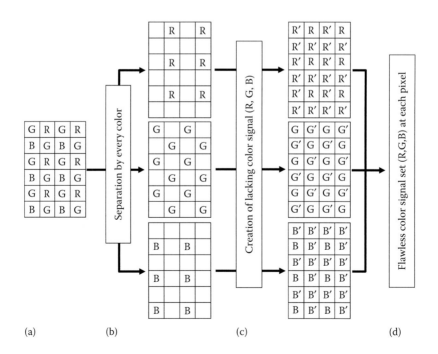

(a)   (b)   (c)   (d)

FIGURE 8.6

Demosaicking (color interpolation): (a) color filter array and pixel signal; (b) separation by every color; (c) color interpolation by color signal creation through demosaicking; (d) completion of RGB set at each pixel.

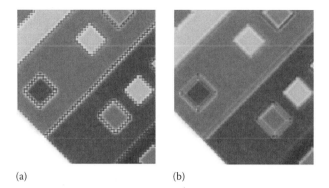

(a)   (b)

FIGURE 8.7 **(See color insert)**
Example of comparison of demosaicking: (a) simple processing: shaggy at color border; (b) adaptive processing: not shaggy at color border.

8.2.4 Color Conversion

Color conversion is signal processing to convert from a color space, which is the signal of sensor response forms, to another corresponding color space in which the color tone is natural for perception by the human eye. Processing is carried out using a linear RGB matrix or color difference matrix. If the sensor signal of the RGB color filter is converted by the RGB matrix, the scale of the signal processing is small.

Denoting the image signal at pixel i after demosaicking, the conversion RGB matrix, and the image signal at pixel i after conversion as $(r, g, b)_i$, A, and $(R, G, B)_i$, respectively, we have the following equation:

$$\begin{bmatrix} R \\ G \\ B \end{bmatrix}_i = A \begin{bmatrix} r \\ g \\ b \end{bmatrix}_i \tag{8.1}$$

$$A = \begin{bmatrix} a_{11} & a_{12} & a_{13} \\ a_{21} & a_{22} & a_{23} \\ a_{31} & a_{32} & a_{33} \end{bmatrix} \tag{8.2}$$

The spectral responses of the color filter and matrix A are decided in a fine balance of color reproducibility with SNR, color resolution, processing load, and so on, as shown in Figure 8.8. While overlapping of each color of the filter is necessary for hue reproducibility, excess overlap increases off-diagonal elements of matrix A to subtract overlap component and thus tends to reduce SNR.

So far, the operation has been implemented in linear processing.

8.2.5 Color and Tone Matching

The image signal is converted to a standard color space suitable for human perception. In practice, it is most commonly converted to a color space named sRGB, which is a global

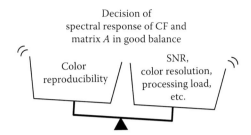

FIGURE 8.8
How to decide color filter response and matrix *A*.

standard adapted to display systems. Digital single lens reflex cameras (DSLRs) often support Adobe RGB, which has a wider color gamut from green to blue, as well as sRGB. Whereas the dynamic range of each color before conversion is 10–16 bit, they should be compressed to 8 bit in sRGB. On that account, color and tone are converted using a nonlinear operation called gamma correction.

8.2.6 Noise Reduction

While correlated noise is efficiently removed by CDS, noises with no correlation or less frequency dependence are reduced by averaging on spatial or temporal axes. Examples of noise reduction by DSP are shown in Figure 8.9.

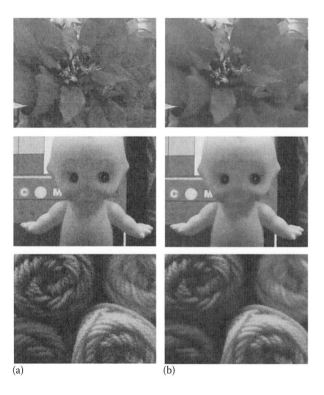

(a)   (b)

FIGURE 8.9   (See color insert)
Examples of noise reduction by DSP: (a) original image; (b) noise-reduced image.

Obviously, color noise is observed in the low signal areas (darker areas) of the original images shown in Figure 8.9a. On the other hand, color noise is seldom observed in noise-reduced images, as shown in Figure 8.9b. However, averaging means degradation of space or time resolution. Therefore, the applicable range has a certain limitation. Readers may have experienced images that have low noise but no sharpness.

8.2.7 Edge Enhancement

To make an image sharper, edge enhancement is often used, which increases the contrast along the boundary to enhance the sharpness. In this way, resolution and sharpness are different. Resolution means how high-frequency (i.e., fine) information can be obtained and is decided by the Nyquist frequency. Sharpness means contrast, in other words, the level difference between white and black, and the signal amplitude, that is, the height of the MTF.

8.2.8 Image Format

Image data output by signal processing are stored in storage media such as memory cards in formats such as JPEG, TIFF and raw data.

8.2.9 Problem of DSP Correction Dependence Syndrome

As mentioned above, DSP is a welcome tool that operates not only essential processing to create image data, but also compensates for the deficiencies of performance sensors, such as inadequate noise reduction. As DSP's performance improves, the area able to be corrected enlarges.

Figure 8.10 shows examples of images processing, including normally captured pictures, pictures processed in the style of paint daubs, and pictures processed in the style of rough

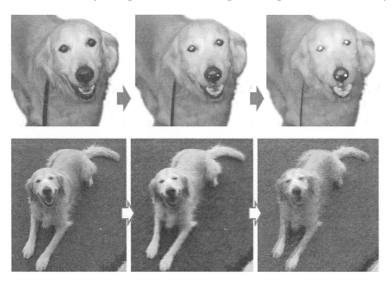

FIGURE 8.10   (See color insert)
Problem of DSP correction dependence syndrome. Where is the border between photography and computer graphics?

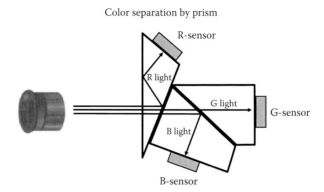

FIGURE 8.11
Schematic diagram of three-sensor camera system.

pastels (top and bottom, respectively). It seems that the border between photography and computer graphics is blurry. While this is allowed for fun, should some kind of guidelines be set for photography?

8.3 Three-Chip Color Camera System

Thus far, descriptions have involved a single-sensor color camera system, which is overwhelmingly dominant in the market because of its low cost and compact system size. Disregarding cost and size, a three-sensor system is the mainstream choice for applications in which picture quality is a high priority. In this system, color separation is implemented by a prism and not by a color filter like in a single-sensor camera system.

As shown in Figure 8.11, light is guided to each of the three sensors assigned as R, G, and B light sensors by reflections at the prism faces. No color filter is set on each sensor. In a single-sensor camera system, only the light that passes through the color filter can contribute as sensitivity in the sensor. The other light is absorbed or reflected by the color filter. However, the color filter's role is to select light. Only one-third of incident light can pass through the color filter and work, making two-thirds unavailable.

As almost all light reaches one of the sensors in a three-sensor camera, there is no such waste. In this system, accurate sensor alignment is necessary so that R, G, and B light from the same portion of the object arrives at the pixel of the same position in each color sensor.

Another advantage of this system is that demosaicking is unnecessary, because R, G, and B light arrives at each pixel in each sensor. As there is no need for signal generation by demosaicking, there is no possibility of it causing false signals, leading to higher picture quality.

Reference

1. T. Kuroda, The 4 dimensions of noise, IEEE International Solid-State Circuits Conference, San Francisco, February 2007, Imaging Forum: Noise in imaging systems, pp. 1–33.

Epilogue

This book was written with the concept of "what is imaging?" as its core. From this perspective, the structure of image information, imaging system architecture, and the impacts of consequent built-in digitized coordinate points in systems have been described.

Technical innovations at the beginning and recent progress in the state of the art have been discussed from the viewpoint of practical use.

As the history shows, techniques actually used are often replaced, for reasons of performance and cost, in the next phase. Limitations of the previous technique and feasibility of the new technique have been shown.

It seems that one of the ultimate targets in the current technology is pixel-level digital signal sensors. However, at a more advanced stage, in the future, we might be released from the "pixel" concept, which provides the benefit of easy signal treatment under the present architecture, but ties us down.

There is much expectation and excitement regarding the functions that will be included in future sensors and it is strongly expected that they will be integrated with brain function sometime in the future.

Takao Kuroda

Index